John William Nystrom

Project of a New System of Arithmetic, Weight, Measure and Coins

John William Nystrom

Project of a New System of Arithmetic, Weight, Measure and Coins

ISBN/EAN: 9783337339586

Printed in Europe, USA, Canada, Australia, Japan

Cover: Foto ©berggeist007 / pixelio.de

More available books at **www.hansebooks.com**

PROJECT

OF A

New System of Arithmetic,

WEIGHT, MEASURE AND COINS,

PROPOSED TO BE CALLED THE

TONAL SYSTEM,

WITH SIXTEEN TO THE BASE.

BY

JOHN W. NYSTROM, C. E.

PHILADELPHIA:
J. B. LIPPINCOTT & CO.
LONDON: TRUBNER & CO.
1862.

Entered according to Act of Congress, in the year 1862, by

JOHN W. NYSTROM,

In the Clerk's Office of the District Court for the Eastern District of Pennsylvania.

KING & BAIRD, PRINTERS, SANSOM ST., PHILADELPHIA.

To the International Association
for obtaining a Uniform Decimal System
of Weights, Measures and Coins:

GENTLEMEN :—

The substance of this book was laid before the meeting held by the British branch of your Association, at Bradford, England, on the 11th of October, 1859. I have since made great effort both in England and America to have the same published, but have not succeeded until now, when done at my own expense. I incur constant outlays for the general interest in promoting science and the arts, for which I expect no remuneration.

Scientific men of good standing have remarked that it would not be proper to publish the correspondence with the Decimal Association, and that with the ———— Society of Philadelphia, but I am of a different opinion. I do not publish the correspondence referred to, for the sake of showing which of us are right or wrong, but simply for the discussion, which I consider to be of some importance in digesting the subject, and it may save others from making similar remarks on the same. The remarks made on my new system of arithmetic, by the Decimal Association, and the ———— Society are of very little importance further than of giving rise to discussion on the subject, but they have proved of great importance in preventing the publication of the herein called *tonal system*.

I shall always be very glad to hear any remarks on the subject, as I am well convinced that the more it is attempted to defend the decimal arithmetic, the more its folly will be exposed.

When the book is published, one copy shall be sent to each member of the International Decimal Association, and I shall feel gratified and compensated if it receives your attention.

The noble object of your Association deserves all possible success, and it is your agreeable duty to endeavour to relieve us of the present dreary and complicated calculations, and to establish such system, as in all its bearings would become the most simple and efficient for all classes of mankind, and our descendants will thank you for ever.

If I knew of no better than the decimal arithmetic, I would get along very well with it, but now unfortunately, when I get entangled in the complication with long decimal tails, I grumble long verses over it.

I have the honor to subscribe myself,

Your most obedient servant,

JOHN W. NYSTROM,

Engineer.

Philadelphia, in January, 1862.

Presidents of the International Decimal Association.

President.

The BARON JAMES DE ROTHSCHILD, Consul-General of the Austrian Empire.

Vice Presidents.

For Belgium...COUNT ARRIVABENE, Brussels.
M. DE BROUCKERE, Burgomaster of Brussels.
J. LIAGRE, Major of Engineers.
A. QUETELET, Astronomer Royal.
J. S. STAS, Professor of Chemistry in the Ecole Militaire, Brussels.
France............ELIE DE BEAUMONT. Senator, Member of the Institute.
MICHEL CHEVALIER, Councilor of State, Member of the Institute.
M. LE PLAY, Councilor of State.
C. L. MATHIEU, Member of the Institute.
GENERAL MORIN, Member of the Institute.
EMILE PEREIRE, Paris.
German Zollverein. DR. STEINBEISS, Privy Councilor, Stuttgart.
Great Britain..HIS GRACE RICHARD WHATELY, D. D., Archbishop of Dublin.
RIGHT HON. THE EARL OF ROSSE, K. P., F. R. S.
RIGHT HON. BARON FORTESCUE.
THE VERY REV. RICHARD DAWES, M. A., Dean of Hereford.
RICHARD COBDEN, Esq., M. P.
JAMES YATES, Esq , M. A., F. R. S.
Greece ....
Holland..........M. VROLIK, formerly Minister of Finance, The Hague.
Italy...............HIS EXCELLENCY THE MARQUIS D'AZEGLIO, London.
CHEVALIER BERTINI, Turin.
SIGNOR BARTOLOMEO CINI, Florence.
CHEVALIER CORRIDI, Director of the Technical Institute, Florence.
SIGNOR ENRICO MAYER, Pisa.
DR. PANTALEONE, Rome.
Liberia............J. J. ROBERTS, Esq., late President of the Republic, Monrovia.
Mexico...........SIGNOR PACHECO, Minister Plenipotentiary at Paris.
Portugal........
M. D'AVILA, Minister of State, formerly Minister of Finance.
JOAQUIM HENRIQUEZ FRADESSO DA SILVEIRA, Superintendent of Weights and Measures, Lisbon.
Russia............MR. KUPFFER, Member of the Imperial Academy of Science, St. Petersburg.
Spain.............SIGNOR RAMON DE LA SAGRA, Member of the Cortes.
Switzerland....PROFESSOR DE LA RIVE, Geneva.
GENERAL DUFOUR, Geneva.
M. TRUMPLER, Zurich.
United States of North America.
DR. A. D. BACHE, Washington.
HON. GEORGE BANCROFT, New York.
HICKSON FIELD, Esq., New York.
HON. CHARLES SUMNER, Boston.

Secretaries.

For England...HENRY COLE, ESQ., South Kensington Museum.
France............H. HIPPOLYTE PEUT, 12 Rue de la Bruyere, Paris.
United States...M. A. VATTEMARE, 17 Rue de Tivoli, Pa

CONTENTS.

	PAGE.
Letter to the International Decimal Association	3
Presidents of the International Decimal Association	5
Introduction	9
Tonal System of Arithmetic	15
Names of the Tonal Figures	16
Tonal and Decimal Numbers compared	18
Tonal and Decimal Fractions	21
Tonal Addition and Multiplication Tables	22
Addition	24
Subtraction and Multiplication	25
Division	26
Tonal Logarithms	27
Tonal Weight, Measure, and Coin	27
Circumference of the Earth	29
Length, Tonal Meter	30
Time and the Circle	32
Tonal Watch, or Clock-dial	34
Tonal Compass	35
Latitude and Longitude, Tonal Division	36
Measure of Surface	36
" Capacity	37
" Weight	37
" Power	38
Money	39
Market Prices	41
Postage Stamps	42

Division of the Year	43
Measure of Heat	44
Music	45
Boem's Flute	50
Abbreviation of Tonal Units	51
Examples for Tonal Calculation	52
Counting Machine for the Tonal System	57
Russian Tschoty	59
Meeting of the International Decimal Association in England	61
Letter from the British Branch of the International Association	62
Letter to the International Decimal Association	65
Critic on the French Meter	75
Verniers, Tonal and Decimal	81
Weights for Weighing	82
Correspondence with the ——— Society of Philadelphia	85
Comments on the Tonal System	87
Reply to the Comment on the Tonal System	89
Binary Division, Discussion on the Term	90
Sixths and Thirds in the Tonal, Decimal and Octonal Systems	91
Prices on Street Railroads	92
Railroad Times	93
Market Prices, Discussion on	94
American Dollar a Medium Unit	95
Market Practice, Discussion on	97
Mitchell's Price Ticket-box	98
Mr. Taylor's Octonal System	101
Tonal, Decimal, and Octonal Systems	102
Calculating Machine, Nystrom's	105
Pocket-Book of Mechanics and Engineering	106

INTRODUCTION.

The introduction of the decimal system of weight, measure, and coins, is steadily progressing in most parts of the world, but when or wherever it is first proposed, it meets with many natural and reasonable objections. The inconvenience of the decimal arithmetic is well known, and better bases for the same have been frequently proposed, but there has not, that I am aware of, been made any earnest attempt, by proper authority, to introduce a system that would in all its bearings constitute the greatest possible simplicity and efficacy, nor to remove the principal objections to the decimal arithmetic. Questions may arise, *first*, what are the difficulties and objections? and *secondly*, how can they be removed and overcome?

The principal difficulty and objection to the decimal system is, that the base 10 does not permit of binary divisions, as required in the shop and the market. In attempting to introduce the decimal system in England, it met the said reasonable objections by Lord Overstone's observation that "the number 12 presents "greater advantage than 10; a coinage founded on "the first number is more convenient for the purpose "of the shop and market." It is evident that 12 is a better number than 10 or 100 as a base, but it admits of only one more binary division than 10, and would, therefore, not come up to the general requirement.

The number 16 admit binary division to an infinite extent, and would, therefore, be the most suitable number as a base for arithmetic, weight, measure, and coins.

The experience of practical men and close observers is, that wherever the decimal system is introduced, it is of more injury or inconvenience to the public than of benefit to the few who have to do with quantities merely by figures on paper.

It is very difficult for the ordinary uneducated classes to understand the decimal system, because they want to divide their things into the most natural fractions, halves, quarters, eighths, sixteenths, &c., &c., for which 10 is not a suitable number.

For calculation on paper, the decimal system is very convenient, when it is not necessary to understand the operation or to impress the value of quantities on the mind, as is the case with many arithmeticians, who manages the figures and comes to the result as easy as a musician who plays the crank-organ. The decimal system is not so easy for the practical man and self-thinker, who impresses the value and relative position of quantities on his mind, as he proceeds in his measurement and calculation. I have, from my early youth, had a great deal to do with different kinds of measurement and calculation, and have always found it inconveniently arranged, not for the different multiplication and division of the units of measures, but principally on account of the arithmetical system not being well planned.

About twelve years ago I invented a calculating machine, which was patented in America in the year 1850; in working out this instrument, a great many ideas suggested themselves as to the improvement of the arithmetical system, weight and measure in general. In a pamphlet printed in Philadelphia in 1851, describing the calculating machine, mention is made about systems of arithmetics with 8 and 16 as the base, with suggestions how to form six new figures for the latter. At that time I acquired a good practice in a system with 16 as the base, and intended to publish in the Journal of the Franklin Institute an article similar to the contents of this, but was not fulfilled.

The base 10 has very likely originated from the 10 fingers on the hands, which latter are even yet used sometimes in counting, but 10 is actually the worst even number that could be selected; 8 or 12 would have been much better, but 16 the very best.

The principal difficulties and objections to the decimal system cannot possibly be overcome without changing the arithmetical base, for which I herein will propose a new system of arithmetic, weight, measure, and coins, with the number 16 as the base, which system will be hereafter spoken of as the *Tonal System*, because its base 10 (16) is proposed to be called *Ton*.

Attempts are now being made in most parts of the civilized world, and an association formed for the purpose of introducing an international decimal system of weights, measures, and coins. It is then just the time to attempt to introduce a better system of arithmetic, which would combine and include *all* requirements of *all* the different classes of mankind. With the present arithmetic, it is utterly impossible to come to a satisfactory decision on a uniform system of weight, measure, and coins.

The International Decimal Association is in favor

of introducing the French metrical system, which is the most complete in existence, but has the evident disadvantages herein alluded to.

In the *tonal system*, what is now generally understood by decimals, will become the most natural for all possible requirement without exception, as well for measurement in the shop and market as for calculation; and particularly so in mental calculation; being based on the binary multiplication and division, makes it most clear to the mind. Such can never be the case with the decimal arithmetic; which will always present its difficulties, and for the same reason the French *metrical system* will never be well received in an English or American machine shop; it is not well suited for practical people. I have myself a great deal of measuring to do, and always prefer the binary multiplication and division. In the machine shop and drawing room I prefer the English foot with inches divided into eighths, but on the ship's floor in laying out lines of vessels I prefer the English foot divided into decimals; a French meter would of course answer the same purpose for the latter. I shall never use a French meter in the machine shop or drawing room, provided I am not obliged to do so by law, and am sure that the majority of English Engineers would say the same.

Should the French *metrical system* be introduced in America and England, the people would of course in time become accustomed to it, and it would always be found to work well by pen and ink, but the binary fractions expressed by decimals would still appear curious to the majority of the people.

It may be remarked that in this age, almost every one has received more or less education, and consequently the decimal system would not present such inconvenience as herein stated—very well—the binary system is as natural for the educated classes as for the uneducated, for which it would be the most natural to introduce an arithmetic that would in all its forms become the most convenient for all parties, and no special education required for the same. It is the *decimal base* which causes the mischiefs and discordance in weight, measure, and coins, while the *tonal base* would create a perfect harmony.

The difficulty of introducing the *tonal system* is more apparent than real. Introduce it first into schools, at the same time it will be picked up by one after the other; when a little practice is acquired, they will soon conceive its utility and simplicity, and encourage others to follow. At the same time, the sixteen new figures with their new names and multiplication table to be published in all almanacs and newspapers; the Governments preparing the new standards for weight, measure, and coins; the watch and clock makers making new time-pieces; the mathematicians preparing their tables of logarithms and trigonometrical lines, &c., &c. The astronomers preparing their tables and almanacs for the land and sea and celestial objects; the topographers altering their maps to suit the new division of the globe; the mathematical instrument makers to alter the angle-measuring instruments and thermometers, all to suit the *tonal system*, and it would soon be complete for introduction. All the different units, multiplied and divided by the base 16, could be introduced and employed with the decimal arithmetic to begin with, when in a few years the *tonal* arithmetic would become most natural with its units.

A few weeks ago I read in the London Engineer for the 25th of March, 1859, an article by the *International Association for obtaining a uniform decimal system of measure, weight, and coins;* also Lord Overstone's remarks in the London Illustrated News; this induced me to resume my old ideas of the arithmetical system with 16 to the base, being on a trip on the river Volga from Tvær to Tsaritzen when, I had plenty of time to devote to this important subject, I worked this out with the intention to submit it to the above mentioned Association.

The *tonal system* herein proposed with six new figures will of course appear strange at the first glance, and may be considered difficult to introduce to the public, but a little reflection will lead to the conviction of its simplicity and importance.

At the end I have given a description of an instrument by which the *tonal system* can easily be acquired, and the mind turned from the decimal base.

On the river Don, Cosack, in May, 1859,

JOHN W. NYSTROM,
Engineer.

PROJECT

FOR A

NEW SYSTEM OF ARITHMETIC

WITH 16 TO THE BASE,

TO BE CALLED THE

TONAL SYSTEM.

In the *Tonal System* it is proposed to add six new figures to the 10 arabic, thus:

1, 2, 3, 4, 5, 6, 7, 8, 5, 9, ?, ?, &, %, ?, 10,

making 16 characters to form the base. In order to form a clear conception of the nature and utility of the *Tonal System*, it will be well to enter into some details of calculation with examples, in connection with which it is necessary to give names to the new figures, or rather to give new names to the 16 characters, so as to clearly distinguish it from our present system.

A new system of this kind could not well be introduced in one country alone, but the whole world at large must agree on its acceptance; it then becomes necessary in the project of the system to select such names of the figures as to make it well suited to all languages, both in spelling and sound; for which the following names are given, without reference to any language or thing. I go so far as to say that I would object to a professor of languages fixing the names of the figures, because he would surely select them from the Hebrew, Greek, or Latin, which, I have no doubt, would make it very pretty, such as *Hecatogramme*, in the French system. It is desirable to have the names clear and simple, in expressing as well compound numbers as the different units for measures. It is not necessary to employ more than one syllable for each object expressed.

The names of the *Tonal figures* are contained in the following four words, *Andetigo, Subyrame, Nikohury, Lapofyton*, which should be learned by heart. The vowel *y* in these names should be pronounced as in the English word *cylinder*, *i* as in *will*, *e* as in *then*, *a* as in *all*.

Tonal Names of Single Figures and Compound Numbers.

0.	Noll.	17.	Tonra.	3𝑓.	Titonfy.
1.	An.	18.	Tonme.	40.	Goton.
2.	De.	1ɔ.	Tonni.	43.	Gotonti.
3.	Ti.	19.	Tonko.	46.	Gotonby.
4.	Go.	1ꝑ.	Tonhu.	4ꞷ.	Gotonhu.
5.	Su.	1ꞓ.	Tonvy.	50.	Suton.
6.	By.	1ɛ.	Tonla.	ꞷ0.	Meton.
7.	Ra.	1ƶ.	Tonpo.	ꞷ0.	Huton.
8.	Me.	1𝑓.	Tonfy.	ꞷꞓ.	Vytonme.
ꞓ.	Ni.	20.	Deton.	ƶ𝑓.	Potonfy.
9.	Ko.	21.	Detonan.	𝑓ɔ.	Fytonni.
ꝑ.	Hu.	22.	Detonde.	100.	San.
ꞓ.	Vy.	24.	Detongo.	101.	Sanan.
ɛ.	La.	26.	Detonby.	102.	Sande.
ƶ.	Po.	28.	Detonme.	106.	Sanby.
𝑓.	Fy.	2ꝑ.	Detonhu.	10ƶ.	Sanpo.
10.	Ton.	2ƶ.	Detonpo.	110.	Santon.
11.	Tonan.	30.	Titon.	11ꝑ.	Santonhu.
12.	Tonde.	31.	Titonan.	120.	Sandeton.
13.	Tonti.	32.	Tidonde.	129.	Sandetonko.
14.	Tongo.	35.	Titonsu.	130.	Santiton.
15.	Tonsu.	39.	Titonko.	145.	Sangotonsu.
16.	Tonby.	3ꝑ.	Titonvy.	200.	Desan.

28ꞇ.	Desan-metonfy.
3ꞓꞌ.	Tisan-latonhu.
700.	Rasan.
ꝛꞇꞌ.	Husan-vytonfy.
1000.	Mill.
2000.	Demill.
8ꝛꞇ5.	Memill-husan-vytonsu.
1,0000.	Bong.
ꞇ,0610.	Vybong, bysauton.
10,0000.	Tonbong.
100,0000.	Sanbong.
1510,0000.	Mill-susanton-bong.
1,0000,0000.	Tam.
1,0000,0000,0000.	Song.
1,0000,0000,0000,0000.	Tran.
2,8ꞓ5ꞇ,7ꞓ0ꞇ.	Detam, memill - lasan - suton - hubong, ramill-posanfy.

This arrangement of expressing numbers is clear and simple, but it requires some practice before the sound impresses the corresponding value on the mind, for which it is necessary to have a clear conception of the sound and value of each figure. The object of employing different consonants to the names of the figures is to render it more difficult to alter a written number from one value to another; it will also make the expression clearer. Although the old figures in the *Tonal System* bears the old value (except 9) one by one, it will not be so in compound numbers, as will be seen in the following table I:

TABLE I.

Notation of Tonal and Decimal Numbers.

Decimal.	Tonal.	Decimal.	Tonal.	Decimal.	Tonal.	Decimal.	Tonal.
1	1	33	21	65	41	97	61
2	2	34	22	66	42	98	62
3	3	35	23	67	43	99	63
4	4	36	24	68	44	100	64
5	5	37	25	69	45	101	65
6	6	38	26	70	46	102	66
7	7	39	27	71	47	103	67
8	8	40	28	72	48	104	68
9	9	41	29	73	49	105	69
10	9	42	29	74	49	106	69
11	↊	43	2↊	75	4↊	107	6↊
12	↋	44	2↋	76	4↋	108	6↋
13	Ɛ	45	2Ɛ	77	4Ɛ	109	6Ɛ
14	↊̃	46	2↊̃	78	4↊̃	110	6↊̃
15	↱	47	2↱	79	4↱	111	6↱
16	10	48	30	80	50	112	70
17	11	49	31	81	51	113	71
18	12	50	32	82	52	114	72
19	13	51	33	83	53	115	73
20	14	52	34	84	54	116	74
21	15	53	35	85	55	117	75
22	16	54	36	86	56	118	76
23	17	55	37	87	57	119	77
24	18	56	38	88	58	120	78
25	19	57	39	89	59	121	79
26	19	58	39	90	59	122	79
27	1↊	59	3↊	91	5↊	123	7↊
28	1↋	60	3↋	92	5↋	124	7↋
29	1Ɛ	61	3Ɛ	93	5Ɛ	125	7Ɛ
30	1↊̃	62	3↊̃	94	5↊̃	126	7↊̃
31	1↱	63	3↱	95	5↱	127	7↱
32	20	64	40	96	60	128	80

TABLE I.

Notation of Tonal and Decimal Numbers.

Decimal.	Tonal.	Decimal.	Tonal.	Decimal.	Tonal.	Decimal.	Tonal.
129	81	161	91	193	ɑ1	225	ꞓ1
130	82	162	92	194	ɑ2	226	ꞓ2
131	83	163	93	195	ɑ3	227	ꞓ3
132	84	164	94	196	ɑ4	228	ꞓ4
133	85	165	95	197	ɑ5	229	ꞓ5
134	86	166	96	198	ɑ6	230	ꞓ6
135	87	167	97	199	ɑ7	231	ꞓ7
136	88	168	98	200	ɑ8	232	ꞓ8
137	8ꞩ	169	9ꞩ	201	ɑꞩ	233	ꞓꞩ
138	89	170	99	202	ɑ9	234	ꞓ9
139	8ƨ	171	9ƨ	203	ɑƨ	235	ꞓƨ
140	8ɛ	172	9ɛ	204	ɑɛ	236	ꞓɛ
141	8ε	173	9ε	205	ɑε	237	ꞓε
142	8ȥ	174	9ȥ	206	ɑȥ	238	ꞓȥ
143	8ƒ	175	9ƒ	207	ɑƒ	239	ꞓƒ
144	ꞩ0	176	ƨ0	208	ε0	240	ƒ0
145	ꞩ1	177	ƨ1	209	ε1	241	ƒ1
146	ꞩ2	178	ƨ2	210	ε2	242	ƒ2
147	ꞩ3	179	ƨ3	211	ε3	243	ƒ3
148	ꞩ4	180	ƨ4	212	ε4	244	ƒ4
149	ꞩ5	181	ƨ5	213	ε5	245	ƒ5
150	ꞩ6	182	ƨ6	214	ε6	246	ƒ6
151	ꞩ7	183	ƨ7	215	ε7	247	ƒ7
152	ꞩ8	184	ƨ8	216	ε8	248	ƒ8
153	ꞩꞩ	185	ƨꞩ	217	εꞩ	249	ƒꞩ
154	ꞩ9	186	ƨ9	218	ε9	250	ƒ9
155	ꞩƨ	187	ƨƨ	219	εƨ	251	ƒƨ
156	ꞩɛ	188	ƨɛ	220	εɛ	252	ƒɛ
157	ꞩε	189	ƨε	221	εε	253	ƒε
158	ꞩȥ	190	ƨȥ	222	εȥ	254	ƒȥ
159	ꞩƒ	191	ƨƒ	223	εƒ	255	ƒƒ
160	90	192	ɑ0	224	ꞓ0	256	100

TABLE II.

Notation of Tonal and Decimal Numbers.

Decimal.	Tonal.	Decimal.	Tonal.	Decimal.	Tonal.
100	64	100,000	1,8690	3,584	꠸00
200	꠸8	200,000	3,0840	3,840	꠵00
300	120	300,000	4,5870	4,096	1000
400	150	400,000	6,1980	8,192	2000
500	184	500,000	7,9120	12,288	3000
600	258	600,000	5,2770	16,384	4000
700	2꠸0	700,000	9,9860	20,480	5000
800	320	800,000	꠵,3500	24,576	6000
900	384	900,000	꠸,꠸꠸90	28,672	7000
1,000	378	1,000,000	꠵,4240	32,678	8000
2,000	780	2,000,000	1꠸,8480	36,864	9000
3,000	꠸꠸8	3,000,000	28,86꠵0	40,960	꠵000
4,000	490	4,000,000	38,0500	45,056	꠵000
5,000	1388	256	100	49,152	꠵000
6,000	1770	512	200	52,348	꠵000
7,000	1758	768	300	57,344	꠵000
8,000	2040	1,024	400	61,440	꠵000
9,000	2308	1,280	500	65,536	1,0000
10,000	2710	1,530	600	262,144	4,0000
20,000	4020	1,792	700	524,288	8,0000
30,000	7550	2,048	800	786,432	꠵,0000
40,000	5840	2,304	500	1,048,576	꠵,0000
50,000	꠸550	2,560	900	16,777,216	10,0000
60,000	꠵960	2,816	꠵00	268,435,456	100,0000
70,000	1,1170	3,072	꠵00	3,489,767,296	1000,0000
80,000	1,3880	3,320	800	55,736,276,736	1,0000,0000
90,000	1,5950				

TABLE III.

Vulgar Fractions, Tonals and Decimals.

Decimal.	Tonal.	Decimal.	Tonal.
$\frac{1}{2} = 0.5$	$\frac{1}{2} = 0.8$	$\frac{11}{16} = 0.6875$	$\frac{2}{10} = 0.2$
$\frac{1}{4} = 0.25$	$\frac{1}{4} = 0.4$	$\frac{13}{16} = 0.8125$	$\frac{3}{10} = 0.3$
$\frac{1}{8} = 0.125$	$\frac{1}{8} = 0.2$	$\frac{15}{16} = 0.9375$	$\frac{7}{10} = 0.7$
$\frac{3}{4} = 0.75$	$\frac{3}{4} = 0.C$	$\frac{1}{32} = 0.03125$	$\frac{1}{20} = 0.08$
$\frac{3}{8} = 0.375$	$\frac{3}{8} = 0.6$	$\frac{7}{24} = 0.29166..$	$\frac{1}{18} = 0.4999..$
$\frac{5}{8} = 0.625$	$\frac{5}{8} = 0.9$	$\frac{5}{12} = 0.4166..$	$\frac{5}{v} = 0.6999..$
$\frac{7}{8} = 0.875$	$\frac{7}{8} = 0.7$	$\frac{1}{3} = 0.3333..$	$\frac{1}{3} = 0.5555..$
$\frac{1}{16} = 0.0625$	$\frac{1}{10} = 0.1$	$\frac{2}{3} = 0.6666..$	$\frac{2}{3} = 0.9999..$
$\frac{3}{16} = 0.1875$	$\frac{3}{10} = 0.3$	$\frac{1}{6} = 0.1666..$	$\frac{1}{6} = 0.2999..$
$\frac{5}{16} = 0.3125$	$\frac{5}{10} = 0.5$	$\frac{1}{64} = 0.015625$	$\frac{1}{40} = 0.04$
$\frac{7}{16} = 0.4375$	$\frac{7}{10} = 0.7$	$\frac{25}{64} = 0.380625$	$\frac{15}{40} = 0.64$
$\frac{9}{16} = 0.5625$	$\frac{5}{10} = 0.5$	$\frac{1}{128} = 0.0078125$	$\frac{1}{80} = 0.02$

TABLE IV.

Addition and Subtraction. Tonal System.

1	2	3	4	5	6	7	8	9	9	?	ℓ	ε	z	ꝼ	10
2	4	5	6	7	8	9	9	?	ℓ	ε	z	ꝼ	10	11	12
3	5	6	7	8	9	9	?	ℓ	ε	z	ꝼ	10	11	12	13
4	6	7	8	9	9	?	ℓ	ε	z	ꝼ	10	11	12	13	14
5	7	8	9	9	?	ℓ	ε	z	ꝼ	10	11	12	13	14	15
6	8	9	9	?	ℓ	ε	z	ꝼ	10	11	12	13	14	15	16
7	9	9	?	ℓ	ε	z	ꝼ	10	11	12	13	14	15	16	17
8	9	?	ℓ	ε	z	ꝼ	10	11	12	13	14	15	16	17	18
9	?	ℓ	ε	z	ꝼ	10	11	12	13	14	15	16	17	18	19
9	ℓ	ε	z	ꝼ	10	11	12	13	14	15	16	17	18	19	1?
?	ε	z	ꝼ	10	11	12	13	14	15	16	17	18	19	1?	1ℓ
ℓ	z	ꝼ	10	11	12	13	14	15	16	17	18	19	1?	1ℓ	1ε
ε	ꝼ	10	11	12	13	14	15	16	17	18	19	1?	1ℓ	1ε	1z
z	10	11	12	13	14	15	16	17	18	19	1?	1ℓ	1ε	1z	1ꝼ
ꝼ	11	12	13	14	15	16	17	18	19	1?	1ℓ	1ε	1z	1ꝼ	20
10	12	13	14	15	16	17	18	19	1?	1ℓ	1ε	1z	1ꝼ	20	

TABLE V.

Multiplication and Division. Tonal System.

1	2	3	4	5	6	7	8	9	9	?	ℓ	ε	z	ꝼ	10
2	4	6	8	9	ℓ	z	10	12	14	16	18	19	1ℓ	1z	20
3	6	9	ℓ	ꝼ	12	15	18	1ℓ	1z	21	24	27	29	2ε	30
4	8	ℓ	10	14	18	1ℓ	1ꝼ	20	24	28	2ℓ	30	34	38	3ℓ	40
5	9	ꝼ	14	19	1ε	23	28	2ε	32	37	3ℓ	41	46	4ℓ	50
6	ℓ	12	18	1ε	24	29	30	36	3ℓ	42	48	4ε	54	59	60
7	z	15	1ℓ	23	29	31	38	3ꝼ	46	4ε	54	5ℓ	62	69	70
8	10	18	20	28	30	38	40	48	50	58	60	68	70	78	80
9	12	1ℓ	24	2ε	36	3ꝼ	48	51	59	63	6ℓ	75	7z	87	90
9	14	1z	28	32	3ℓ	46	50	59	64	6z	78	82	8ℓ	96	90
?	16	21	2ℓ	37	42	4ε	58	63	6z	75	48	8ꝼ	59	95	?0
ℓ	18	24	30	3ℓ	48	54	60	6ℓ	78	84	50	3ℓ	98	ℓ4	ℓ0
ε	19	27	34	41	4z	5ℓ	68	75	82	8f	5ℓ	99	?6	ℓ3	ε0
z	1ℓ	29	38	46	54	62	70	7z	8ℓ	59	98	ℓ6	ℓ4	ε2	z0
ꝼ	1z	23	3ℓ	4ℓ	59	69	78	87	96	95	ℓ4	ℓℓ	ε2	z1	ꝼ0
10	20	30	40	50	60	70	80	90	90	?0	ℓ0	ε0	z0	ꝼ0	100

EXPLANATION OF TABLES.

TABLE I. shows the different notation of equal numbers in the *decimal* and *tonal* systems, where it will be seen that the new system require a less number of figures in expressing a high number; *decimal* 134 = 86 *tonal*, yet the real value is the same in both cases.

TABLE II. is a further extension of Table I. useful for transferring numbers from one system to the other.

EXAMPLE 1. Required how the number 31,868 will be noted by the *tonal system?*

$$\text{Decimal.} \left\{ \begin{array}{r} 30,000 = 7550 \\ 1,000 = 3\tilde{o}8 \\ 800 = 320 \\ 68 = 44 \\ \hline 3,1868 = 7\text{b}3\text{c} \end{array} \right\} \text{Tonal.}$$

EXAMPLE 2. The year 1859 expressed by the *tonal system*, will be 739, or it would apparently carry us back over 11 centuries.

EXAMPLE 3. A lady of 35 years, required how old she will be by the *tonal system?* The answer is 23 years.

TABLE III. is an excellent illustration of the utility of the *tonal system*. It contains the ordinary fractions used in the shop and the market. It will be seen that the vulgar fractions in daily use, require four to seven decimals, where the *tonal system* require only one or two figures. It must be admitted that it is more natural to divide things into halves, quarters, eighths or sixteenths, than into fifths or tenths, and when the natural fractions are expressed by decimals, they become too complicated for the ordinary uneducated mind, as $\frac{3}{16}$ is equal to 0.1875, which I am assured, cannot be conceived by the very best arithmeticians, but they know by practice in calculation that it is so. In the *tonal system* it is very easy to conceive that $\frac{3}{16}$ is equal to 0.3

TABLE IV. is for addition and subtraction, arranged in the ordinary way, that where the vertical and horizontal columns cross one another is the sum of the index numbers.

EXAMPLE 4. $3+5=8$, $5+9=\mathcal{T}$, and $\mathcal{V}+\mathcal{Z}=15$.

For subtraction, find the greatest number in the column in which the smaller number is the index, and the index of the cross column is the difference, as $17-\mathcal{V}=\mathcal{E}$.

ADDITION.

Ex. 5. $\begin{cases} \text{To} & 36\mathcal{Z}9\mathcal{Z} \\ \text{Add} & 10\mathcal{O}7\text{S} \\ \hline \text{Same} & 47526 \end{cases}$ $\left.\begin{array}{l} \mathcal{Z}+8=16 \\ 9+7=11 \, . \\ \mathcal{V}+\mathcal{O}=14 \, . \, . \\ 6+0=6 \, . \, . \, . \\ 3+1=4 \, . \, . \, . \, . \\ \hline 47526 \end{array}\right\}$

Ex. 6. $\begin{cases} 89\mathcal{T}\mathcal{O} \\ 45\mathcal{Z}\mathcal{V} \\ 30\text{S} \\ \hline \mathcal{E}3\mathcal{Z}\mathcal{V} \end{cases}$ $\left.\begin{array}{l} \mathcal{O}+\mathcal{V}+8=1\mathcal{V} \\ \mathcal{T}+\mathcal{Z}+0=1\mathcal{E} \\ 9+5+3=12 \\ 4+8=\mathcal{V} \\ \hline \mathcal{E}3\mathcal{Z}\mathcal{V} \end{array}\right\}$

Ex. 7. $\left\{\begin{array}{r}3519·8ᘔ\\ 6ᘔ0·01\\ ᘔ3·31\\ 0·03\\ 0·49\\ \hline 3ᘔᘰ8·61\end{array}\right.$ Ex. 8. $\left\{\begin{array}{r}67ᘔ50\\ ᘔᘰ0ᘰ\\ 915\\ 9ᘰ\\ ᘔ\\ \hline 7405ᘔ\end{array}\right.$

SUBTRACTION.

Ex. 5. $\left\{\begin{array}{l}\text{From } 3ᘔᘰ9ᘰ\\ \text{Subt. } 4ᘔ53\\ \hline \text{Diff. } 3431ᘰ\end{array}\right\}$

Ex. 9. $\left\{\begin{array}{r}+\ 8104ᘔᘰ\\ -\ 4250ᘰ\\ \hline 7ᘰ8ᘰᘔᘰ\end{array}\right.$ Ex. ᘔ. $\left\{\begin{array}{r}+\ ᘔ9ᘰ80·01ᘰ\\ -\ ᘔ00ᘰ·301\\ \hline 7ᘰᘔᘰ0·8ᘔᘔ\end{array}\right.$

In all arithmetical operations, the *tonal fractions* work precisely the same as *decimal fractions*.

TABLE V is an ordinary arranged multiplication table.

MULTIPLICATION.

Ex. ᘰ. $\left\{\begin{array}{r}38926\\ 6\\ \hline 154044\end{array}\right.$ $\begin{array}{l}6\times 6=24\\ 6\times ᘰ=42.\\ 6\times 9=3ᘰ..\\ 6\times 8=30...\\ 6\times 3=12....\\ \hline 154044\end{array}$

Ex. ᘔ. $\left\{\begin{array}{r}805ᘰ9\\ 72\\ \hline 1013ᘰ4\\ 384586\\ \hline 3547154\end{array}\right.$ Ex. ᘔ. $\left\{\begin{array}{r}38ᘔ706·4ᘰ\\ 0·00684\\ \hline ᘔ2ᘰ8153ᘰ\\ 1ᘰ5ᘰ03278.\\ 1547125ᘔ9..\\ \hline 171ᘔ·6211950\end{array}\right.$

2

DIVISION.

Ex. 9.

3 | 1890970 | 835519·99
18 . . . . .

09 . . . .
5 . . . .

10 . . .
18 . . .

18 . .
18 . .

09 .
9 .

20
18

20
18

20

Ex. 10.

182 | 40895008 | 290065·13393
394 . . . . .

1249 . . . .
1234 . . . .

1650 . .
1588 . .

980 .
970 .

548
519

230
182

570
576

690
576

1290
1234

690

Table of Tonal Logarithms.

Number.	Logarithm.	Number.	Logarithm.
1	0·0	5	0·ƐƄ
2	0·4	9	0·Ƅ4
3	0·66	Ɛ	0·ƐƐ
4	0·8	Ɛ	0·Ɛ6
5	0·55	Ɛ	0·ƐƐ
6	0·96	Ƅ	0·Ƥ4
7	0·Ƅ4	Ƥ	0·ƤƐ
8	0·Ɛ	10	1·00

This table of *tonal logarithms* is a good illustration of the simplicity of the system. In logarithms for single figures, the montissa contains only one or two *tonals*, where the decimal system has a tail of an endless number of decimals.

TONAL SYSTEM OF WEIGHTS, MEASURES, AND COINS.

A unit for the measurement of length ought to be of a convenient size for the artizan in laying out work. Units of about the length of the English foot seems to be almost universally adopted, but it will be observed that the artizan generally employs two such units, or a two foot rule, which length appears to be the best suitable for the actual operation of measurement. In some countries *units* of about this length are employed, as the Swedish *aln*, Russian *archin*, the *elle* of Germany, and others, most of them approaching the length of about 2 feet or the footstep of a man. The French *meter* is the longest unit employed for ordinary measurement in the shop and the market. In accordance with my own observation on the actual performance of laying out work with the French *meter*, and also from comments made by Frenchmen, I believe that the meter is *too long*, to be convenient for the artizan, and when made for the pocket, a great many joints are required, which are objectionable in its application. The *meter* being divided into 100 parts makes it inconvenient to divide the joints, ten are too many,—four will contain the odd number 25 centimeters or $2\frac{1}{2}$ decimeter in each part,—five parts are not practical. When the artizan applies the *meter*, he cannot well see the correctness at both ends without placing himself in an inconvenient position, by which the correctness of the measurement is liable to error,— having myself in Paris been witness to the fact alluded to. The French have divided the quadrant of the earth into fixed number 10,000,000 parts in preference to giving the artizan a convenient unit for his measurement. The division of the quadrant of the earth is merely once a matter of calculation, and could easily be divided into an odd number, rather than to give the artizan a unit which does not suit him. If the 10,000,000 parts had some even relation with the general division of the earth's great circle, as to the length of one degree or minute. it would have furnished a good reason for the length of the meter. The quadrant of the earth divided into the most natural or binary divisions *halfs* and halfs, would lately arrive to a length of about $23\frac{1}{2}$ inches, which would have been a much more suitable unit than the *meter* which is nearly 40 inches.

When a new unit of length is to be selected, it ought to be so adjusted as to bear an even relation to the length of minutes and seconds on the great circle of the earth. By the *tonal system* it would become the most natural to divide the circle of the earth repeatedly by the *tonal base*. The mean circumference of the earth is about 24851·64 miles, or 131216659·2 feet, which latter divided by $16 \times 16 \times 16 \times 16 \times 16 \times 16 \times 16 = 16^7$ (1000,0000 tonal) would be 3,489,767,296 parts, each of a length of 0·48882 feet, or 5·86584 inches. Suppose this to be adopted as a unit for the measurement of length, and to be divided and multiplied by the *tonal base*, its full size appearance will be as shown by fig. 1. (This figure is drawn on the rule fig. 3). *Meter* seems to be a good and proper name for the unit of length, and will therefore retain that word by calling the new unit the *tonal meter*.

It seems to me that the most correct way of ascertaining the size of our globe, would be to measure the longest straight distance on the land, which by examining the map, we will find is in about 31° north latitude; starting from Changhae China, through Tchintou, over the Himalaya mountains, Bassora, Isthmus of Suez, Cairo, Cadames, to Santa Cruz, a distance of about 7700 miles, or over 130 degrees in longitude. This distance should be ascertained by actual measurement, compared with astronomical observations, and fixed points located at every *timton*. The only obstructions, though not serious, in this distance, are, lake Tai-Hou in China, about 40 miles, and the Himalaya mountains.

The same could be repeated in America, from Washington to San Francisco, in the 38th parallel, a distance of about 2240 miles, or about 45 degrees difference in longitude. When these distances are known with their corresponding latitude and longitude, the great mean circumference of our globe is easily calculated by well known rules in mathematics. Every Nation could by the same rule find out the length of the standard *tonal meter* in their own country.

Length.

Tonal System.		Old System.
$\frac{1}{1000}$ m.	= 1 Metermill	= 0·001432 inches Eng.
$\frac{1}{100}$ m.	= 1 Metersan	= 0·022913 "
$\frac{1}{10}$ m.	= 1 Meterton	= 0·366615 "
ONE METER		= 5·86584 in.=0·48882 ft.
10 meter	= 1 Tonmeter	= 7·82112 feet.
100 meter	= 1 Sanmeter	= 125·135 "
1000 meter	= 1 Millmeter	= 2002·207 "

It will be perceived that when the word meter is placed before the expression of value, it impresses on the mind a fraction, as $meter\text{-}mill = \frac{meter}{mill}$ or $\frac{1}{1000}$ of a meter; and when the expression of value is placed before the unit, it denotes a multiplication of the same, as Tonmeter = 10 meter.

The minute measurements, as wire and needle gauges, to be tonally numbered and divided. In the present Birmingham wire gauge the highest number denotes the smallest dimension, which ought to be the reverse. Suppose the *meterton* to be divided into 100 *tonal* parts (256 decimal) or metermills, each would be about ¼ of No. 36 B. W. gauge, or 0·001432 of an inch; this part to be noted No. 1 and the *meterton* would be No. 100 which is about $\frac{3}{8}$ of an inch. By such arrangement, the very expression of the number impresses the mind of the real size of the minute measure, derived from the main standard, the circumference of the earth

Such a gauge would most likely be generally adopted for minute measurements in shops where the present B. W. gauge was never known.

The *tonal meter* to be employed in manufactories, for measuring machinery, &c., corresponding to the English foot. The artizan generally carries a two foot rule, folded into two or four parts, the *tonal* measure would be of precisely the same shape, but with four units instead of two.

On the accompanying plate are full size drawings of the *tonal measure*, of which fig. 2 is a four-folded rule of one *meter* in each part, in appearance very much like ordinary four-folded two foot rule. The side A A contains the *meter* tonally divided and numbered. The other side B B of the same rule, contains divisions for circumference and areas of circles, arranged so that opposite the diameter on A is the circumference on B and area on C. Suppose the diameter to be 1·76 meters on A, it corresponds with 5·48 meters the circumference on B, and 2·5 square meters the area on C. The small divisions between B and C are each *four meter-sans* drawn from A to assist the transference and reading on B and C.

Fig. 3 represents a two-folded tonal measure, similar to the English two-folded two foot sliding rule, numbered and divided same as fig. 2. The part E on which fig. 1 is drawn, to receive numbers of specific gravity of substances, and other co-efficients of general use in practice. The scales F, G, and H, are the ordinary sliding rule, divided into the *tonal system*, which in this case stands in such relation to the divisions on the side D D, that any number on H, corresponds with its logarithm on D. The operation on the *tonal* slide rule will be the same as that on the ordinary decimal one.*

The clear and simple relation between numbers and logarithm in the *tonal* system has led me to some valuable conclusions in reference to calculating machines, and mathematical instruments, which I believe would be of the greatest service to the world.

The *Tonmeter*, (7·82112 feet) to correspond with the *Fathom*, to be used for measuring ropes, cables, depths of water, &c., &c. The *Sanmeter* (125·135 feet) to be the length of the surveying chain, to consist of 100 (256 decimal) links of one *meter* each.

The *Millmeter* (2002·207 feet) for road measure and distances at sea, to correspond with miles. One *millmeter* is equal to one *timmill*, see division of time. Longer distances on the earth's surface would be expressed in *Tims*.

Astronomical distances would be best to express in great circles of the globe, by which the mean distance to the sun would be 7·f1 circles. Great distances, such as to fixed stars, could be easier conceived by this measure.

Time and the Circle.

	Tonal System.	Old System.	
One circle	= 10 Tims	= 24 h'rs or 360 degrees.	
1 Tim	= 10 timtons	= 1½ "	22° 30'
1 timton	= 10 timsans	= 5$_{\square}^m$ 37½^s·	1° 24' 22½"
1 timsan	= 10 timmills	= 21·1^s·	5' 9"
1 timmill	= 1 Millmeter	= 1·31836^{seconds}·	19·77"

The length of a pendulum vibrating *timmills* will be 8·555 meters = 67·975 inches.

---

* A few tonal measures are now being made in Philadelphia.

The time, circle, and compass would thus be equally divided, and greatly simplify all astronomical and nautical tables and calculations.

In expressing time, angle of a circle, or points on the compass, the unit *tim* should be noted as integer, and parts thereof as *tonal fractions*, as 5·86 *tims* is five times and *metonby*. The unit *tim* to be pronounced as in the English word *timber*.

The accompanying figures are drawings of a clock or watch dial, and a compass on the *tonal system*.

Fig. 4 shows the appearance of a *tonal* clock dial, the time indicated is 9·3*l*, which expressed by words should be *Kotim* and *titonhu*. The tim hand goes round only once in a night and day, being on 0 at midnight, and on 8 at noon. If a third index hand is added on the same centre, to represent the second hand on our ordinary watch, it should make one turn on the dial for each timsan, when the small division on the circle would indicate *limbonys* or $\frac{1}{10000}$ part of the *tim*, which is $\frac{82}{1000}$ parts of our present second. Such delicate measures of time are often required in Physical Science, as in Astronomical observations, velocity of light and electricity, gunnery, &c. A further extension of delicate measures of time will be conceived in musical vibration, which I shall arrange into immediate connection with the *tonal* watch, that the base note for the natural key, shall make 10 (16 dec.) vibrations per timmill. Turn yourself towards the south with the *tonal* watch in your hand, and it will be found that the timhand follows the sun nearly; or lay the Watch horizontally, so that the timhand points towards the sun, and the figures on the dial will give 0 north, 8 south, 4 east and *t* west, nearly.

Fig. 4.

This is easily comprehended by the public, as the *tonal* compass, fig. 5, is divided the same way. A course noted from the *tonal* compass is clear and simple.

One *millmeter* in length on the equator corresponds with one *timmill* in time. By this division of time, it is always clear whether it is in the morning or evening, without any special notation. Our present system often leads to error or confusion, whether a noted time is meant in the morning or evening.

Fig. 5.

Division of the Earth's great Circle.

The latitude or meridians should be divided from north to south into 8 *tims*, with 0 at the north pole, 4 *tims* at the equator, and 8 at the south pole. The equator to be divided same as the clock or compass.

Nations ought to agree, to count the longitude from one meridian drawn through a fixed point on the globe. The different notation of longitude on maps is a great inconvenience and sometimes causes confusion. In my present traveling I have maps on which the longitude is noted, some from Greenwich, some from Paris, Pulkova, Washington, Ferro, and on some maps it is not stated from where the longitude is counted. Independently of the different points from where the meridians are noted on maps, the present divisions of the circle make it very complicated to calculate the difference of time between places, and very few will understand how,—in fact the complication is such as to discourage many persons from the attempt; while, if the circle and time were divided as herein proposed, the very figures denoting the meridians would give the difference of time by simple subtraction.

In the Canary Islands appears to be a proper point to place the zero-meridian, as the ancient geographers who have taken their first meridian from the west side of the Island of Ferro 17° 52′ west from Greenwich.

Maps constructed on such principle, would to our descendants forever indicate, not only the true position of the place on our globe, but the scale of latitude would give all distances on the maps in miles, (timmills) feet (meters) and inches (metertons) also the area in acres; and a difference of latitude placed along a parallel, would give the correct distance corresponding with time in longitude. Those plain matters are by our present system, not only complicated in calculation, but are seldom thought of, for the complication screens away the simple knowledge.

Measure of Surface.

Tonal System.		Old System.
One square meter	= 0·239	square feet.
1 Square tonmeter	= 61·15	"
1 Square sanmeter	= 15658·768	"
1000 Square meters	= 0·36 Akres.	

The square sanmeter to be the measure of land, corresponding to the acre.

Measure of Capacity.

Tonal System.		Old System.
1 Gallsan = 10 Gallmills	=	0·79 cub. in. about a table spoon.
1 Gallton = 10 Gallsans	=	12·62 cub. in. about a tumbler.
1 Gall = 1 Cub. Meter	=	201·78 cub. in. about a gallon.
1 Tongall = 10 Galls	=	1½ Bushel.
1 Sangall = 10 Tongalls	=	about 30 cub. feet.
1 Millgall = 10 Sangalls	=	478·2 cubic feet.
1 Millgall = 1 cub. tonmeter	=	17·75 cubic yards.

The *Gall* or *cubic meter* to be the unit for measures of capacity, in ordinary market practice. The *Sangall* to be the measure of excavation and embankments, also for grain, &c. The *Millgall* to be the measure of firewood, being one *cubic tonmeter*.

Measure of Weight.

One *cubic meter* of distilled water will weigh 7·30174 pounds avoirdupois, to be the tonal unit for weights, and to be called a *Pon*.

Tonal System.		Old System.
1 Ponmill		= 0·0_848 drams avoi.
1 Ponsan = 10 Ponmills		= 0·45568 " "
1 Ponton = 10 Ponsans		= 0·45568 pounds "
1 Pon = 10 Pontons		= 7·3017 " "
1 Tonpon = 10 Pons		= 116·8 " "
1 Sanpon = 10 Tonpons		= 1868·8 lbs.= 0·838 tons.
1 Millpon = 10 Sanpons		= 13·34 tons.

The pressure of the atmosphere will be about 46 *pons* per *square meter*, and the height of a column of mercury balancing the atmosphere, about 5 *meters*.

The force of gravity will cause a body to fall 35·27 meters in the first timmill in a vacuum, and the end velocity will be 72·56͡ meters per timmill.

The *Ponsan* to be the unit for apothecary and minute weights. *Pon* for the ordinary market practice. *Sanpon* as shipping unit and heavy weights, corresponding with the ton.

Measure of Power.

One *pon* lifted one *meter* in one *timmill*, to be called one *effect*. By the present system, *one pound lifted one foot in one second is called one effect*, of which there are 550 effects on one horse-power or 55 effects on one man's-power.

	Tonal System.	Old System.
One effect		= 2·704 effects.
1 man's-power = 10 effects	= 43·268 eff.	= 0·86 man.
1 horse-power = 10 men	= 692·3 eff.	= 1·25 horses.

The *man's-power* to be the unit for manual labour, and *horse-power* for machinery and heavy work.

Money.

The American dollar is nearly the mean difference of all the monetary units of the world, and curious enough, compared with the largest the English pound sterling £, and the smallest, the French Franc F, the Dollar D, will be the mean proportion of the two.

$$L : D = D : F. \text{ or } D = \sqrt{LF}.$$

If the world could agree to adopt one unit for money, it seems that the dollars has a claim to be chosen as a standard.

Tonal System.	Old System.
One dollar = 10 shillings	One dollar = 100 cent.
1 shilling = 10 cents	= $6\frac{1}{4}$ cents.
1 cent	= 0·39 cent = 2 centims.

The inconvenience of the monetary decimal system is daily felt in the actual market practic, for although the dollar is divided into 100 parts, for which suitable coins (most of odd numbers of dollars and cents) are in circulation, the retail prices of most articles are fixed to suit the dollar divided into 16 parts. A drink of almost any description costs 6 or $6\frac{1}{4}$ cents, which is $\frac{1}{16}$ part of a dollar. A ride in an omnibus costs generally 6 cents. Suppose an article bought for 6 cents, and paid with a quarter of a dollar, there will be 19 cents change, summed up by the following coins 10ct. + 5ct. + 3ct. + 1ct. = 19 cents. This can reasonably be called an odd system of calculation, because there is nothing but oddity about it. By the *tonal* money, the same article paid by a quarter, which would be 4 shillings, there would be 3 shillings change, in which transaction the mind was carried only to 4, while the decimal money was fumbling about among the odd numbers up to 25.

	Tonal Coins.		United States Coin
Copper,	1 cent. 2 cents. 4 cents.	Copper,	1 cent
Silver,	8 cents. 1 shilling. 2 shilling. 4 shilling. 8 shilling. 1 dollar.	Silver,	3 cents. 5 cents. 10 cents. 15 cents. 20 cents. 25 cents. 50 cents. 100 cents.
Gold,	1 dollar. 2 dollars. 4 dollars. 8 dollars. 10 dollars. 20 dollars.	Gold,	1 dollar. 3 dollars. $2\frac{1}{2}$ dollars. 5 dollars. 10 dollars. 20 dollars.

The *tonal coins* are all of even and of the easiest countable numbers, such as are required in the market, while the *decimal coins* are most of odd numbers, and of a complicated composition for calculation, even the half dollar or 50 cent has a prime number to its index. The *tonal coins* give a nicety of $\frac{1}{256}$th part of a dollar, while the *decimal coins* give it only to $\frac{1}{100}$ part.

The legal interest on money, in most countries is about 6 per cent, which by the *tonal* system would be nearly 10 per sant; consequently, calculating that interest on money, would be only to point off two figures on the capital.

If the *tonal* interest is 1 more or less than 10 per sant, it is calculated by simple addition or subtraction.

Interest on 32534 dollars at 10 per sant. = 325·34
" " 1 " = 35·534
" " 11 " = 3T0·784

which is 3T0 dollars, 7 shillings, and $8\tfrac{1}{4}$ cents.

This makes a simple interest calculation in the neighbourhood where it is most wanted. The difference of 1 per cent interest in the neighborhood of 6, is a rather large margin, for which we often find it accompanied with a fraction, in practice. One per sant *tonal* = 0·391 per cent decimal. One decimal per cent. is 2·56 *per sant tonal*. Fractions would be rarely required to the percentage in the tonal system.

The most common retail prices of articles in America are as follow:

Market Prices ct.	Tonal Shillings or 10ths of a Dollar.	Nearest Decimal Cents.
$6\tfrac{1}{4}$	1	6
$12\tfrac{1}{2}$	2	12 or 13
$18\tfrac{3}{4}$	3	19
25	4	25
$31\tfrac{1}{4}$	5	31
$37\tfrac{1}{2}$	6	37 or 38
$43\tfrac{3}{4}$	7	44
50	8	50
$56\tfrac{1}{4}$	9	56
$62\tfrac{1}{2}$	9	62 or 63
$68\tfrac{3}{4}$	ᚅ	69
75	ᚅ	75
$81\tfrac{1}{4}$	ℰ	81
$87\tfrac{1}{2}$	⸱	87 or 88
$93\tfrac{3}{4}$	ᚖ	94
100	10	100

It may be remarked that those prices are retained from the circulation of Spanish Coins in the United States, to which I beg to reply that if such prices and coins were not the most natural to the mind, and the most suitable for the market they would not be retained.

Postage Stamps.

The following are the Post stamps of the United States.

1ct., 3ct., 10ct., 12ct., 24ct., 30ct., 90ct.

The very first glance at this series shows plainly that there is some confusion about it. The stamps of even post prices are not even in a dollar (except 1ct.) and four of them are not even in any coin, there is a 10 cent post stamp and no 10 cent postage. The simple and even numbers, most valuable in calculation, as 2, 4, 8 and 16, are of necessity omitted, because the decimal system does not admit the natural numbers. Let us now turn our attention to

Tonal Post Stamps.

Tonal Stamps.		American Cents.
4 cents	for city post	$= 1\frac{9}{16}$ cents.
8 cents	" single letters	$= 3\frac{1}{8}$ "
1 shilling	" double letters	$= 6\frac{1}{4}$ "
2 shillings	" quadruple letters	$= 12\frac{1}{2}$ "
4 shillings	" 8 " "	$= 25$ "
8 shillings	" 10 " "	$= 50$ "
1 dollar	" 20 " "	$= 1$ dollar.

Here it will be found that the *tonal post stamp series*

contains the even number most simple for calculation, and they are even both in post prices and in the *tonal* coins or dollar.

Division of the Year.

The new year ought to commence at Christmas, and the year divided into 10 (16 deci.) months, which would make about 17 (23 deci.) days per month.

Seasons	Number of the month	Number of days per month	Names of the new months.	The first day of the new month to commence on.	
Winter	1	17	Anuary.	21 December.	New year and Christmas.
	2	16*	Debrian.	13 January.	
	3	17	Timander.	4 February.	
Spring	4	16	Gostus.	27 February.	Night and day of equal length.
	5	17	Suvenary.	21 March.	
	6	17	Bylian.	13 April.	
	7	17	Ratamber.	6 May.	
Summer	8	17	Mesudius.	29 May.	Midsummer day, or St. John.
	ö	17	Nictoary.	21 June.	
	9	17	Kolumbian.	14 July.	
	ℓ	17	Husamber.	6 August.	
Autumn	ℓ	17	Vyctorious.	29 August.	Night and day of equal length.
	ε	16	Lamboary.	21 September.	
	ŏ	17	Polian.	13 October.	
Winter	ƒ	17	Fylander.	5 November.	
	10	17	Tonborius.	28 November.	

* 17 Days in Leap Year.

There will be 168 tonal days in a year, and 16ŏ in leap years.

The names of the *tonal months* are given so, that the first syllable expresses the number of the month in the year, and every four months have a similarity in sound, impressing the quarters of the year. The new year and Christmas should be on the same day. There is no occasion for altering the days in the week, but when days are to be expressed by *tonal fractions* of the months or year, the number of days are nearly 80 per sant more than the fraction, for instance 6 days = 0·4 months or 0·04 years, 3 days = 0·6 months, 0 days = 0·8 months, 𝔶 days = 0·09 years, 12 days = 0·0𝔶 years, &c., &c. The different Kalenders used in different Countries, would by the *tonal system* at once fall into *one*. The old or Julian style is yet used in Russia and other Countries, it is 12 days behind our new or Gregorian style. The Evangelistic year commences December 27, on the day of St. John; this style is also adopted in Freemasonry, where it is known as the Masonic year. The *tonal style* would become seven days ahead of the Gregorian.

Measure of Heat.

The three different thermometrical scales causes a great deal of inconvenience in science and art. Although Fahrenheit's scale is generally employed in the United States, yet we have American Scientific books in which Celcius' scale is used exclusively. Celcius' decimal scale is the most convenient for calculation, but I believe that those degrees are too large for scientific purposes, that we want the scale to be divided into more parts between the freezing and boiling points.

By the *tonal* system it would become the most natural to divide the thermometer scale into 100 (256 decimal) parts between the freezing and boiling points of fresh water.

Tonal System.		Old System.
Zero or 0	$= +32$	Fah. or 0 Celcius.
1 Temp = 10 tempton	$= 11\frac{1}{4}$	Fah. or $6\frac{1}{4}$ "
1 Tempton	$= 0.7$	Fah. or 0·4 "

Temp for rough measurement of the temperature of the weather, and *tempton* for scientific purposes.

Music.

The many different clefs used in music, seems to be a complication without remuneration. Music Corps often use four different clefs, namely, *Bass*, *Tenor*, *Treble* and *Alto*, all of which could be of one single denomination. The *Tenor* and *Alto* clefs are gradually withdrawn, but there is yet no indication of dispensing with the Bass or F. clef. In Piano music particularly, it is an unnecessary complication, and burdens the student, to have a different denomination on each stave. I shall here arrange it so that all the different clefs will be represented in one denomination.

The standard pitch of tone, I will assume to 100 (256 Arabic) vibrations per timmill, for the base note in the natural key. As A is the first letter in the alphabet, it appears natural that it should be the first or base note in the natural key, and that such an octave from A to A, should be located within the musical stave.

I would propose to denote five different clefs, as follows:

CANTO CLEF. *Soprano.*—This clef to run two octaves above the treble pitch.

ALTO CELF. *Contralto.*—One octave above the treble, for high female voice.

TREBLE CLEF. *Descant.*—Natural position, or ordinary female voice.

TENOR CLEF. For the common voice of man, one octave below the treble.

BASS CLEF. The ordinary bass, two octaves below the treble pitch.

In running music, where the notes extend too high above, or too deep below, the stave, the octave can be altered, by placing the suitable clef on the bar where the change is required, similar to that now employed as 8".

The following natural scales, in the different clefs, show the number of standard vibrations per timmill, of each note, divided according to the geometrical progression or tempered scale, as employed in practical music:

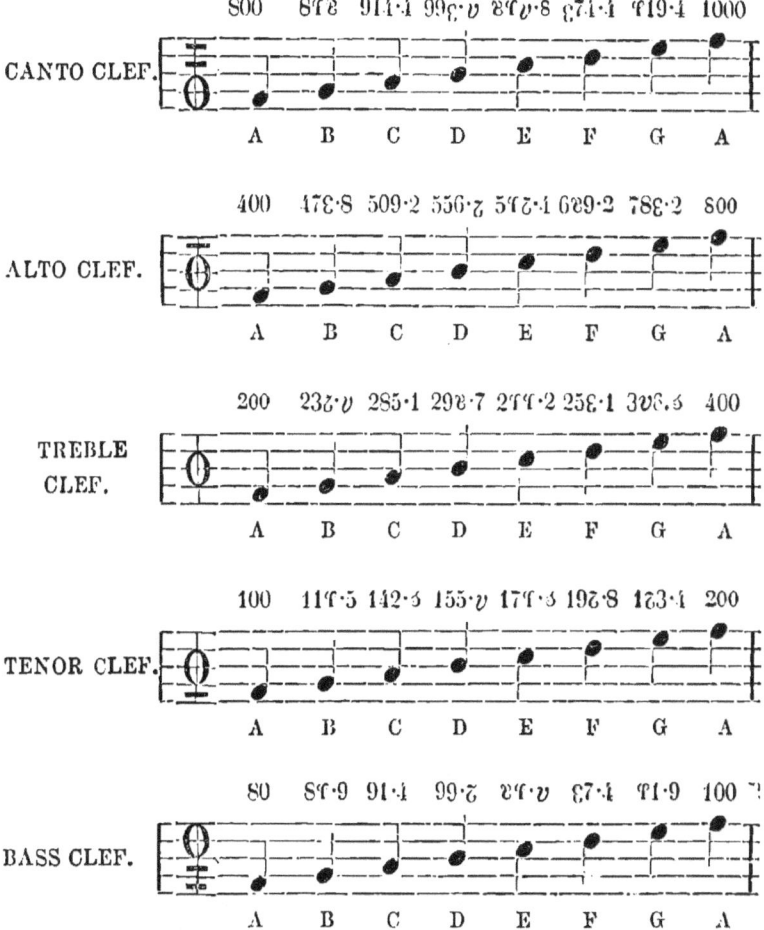

The natural key-note A, vibrating 100 (256 Arabic) per timmill, will be 194 per second, which corresponds nearly with F♯ in the high pitch now used. Complaints have often been made, that our present pitch of music is too high, and it has been proposed to lower it down to 256 vibrations per second of C, as accepted in acoustics, when the tonal pitch of A would correspond nearly with the present G, by which the comparative cromatic scales would be as follows:

By this arrangement, the present key of G should be the natural *tonal* key. The notes on the musical stave would, in the *tonal* system, bear the same names as in the old Bass clef; but the distance between C and D would be only half a note, also that between G and A, as shown on the tonal scale.

The advantage of this arrangement would be to attain a universal standard pitch, and to have only one simple denomination in all music.

Examining the nature of fingering on most musical instruments, we will find that the key of G is most naturally located, as in the Flute, Clarionet, Violin, Guitar, and many others.

Brass instruments ought to be arranged so as to give the natural chord, without using the keys or pistons.

The key-board on the Piano should be moved down five half notes, for the same strings to suit the *tonal* music.

The different keys in sharps and flats would appear as follows:

Of all the different sciences, that of music seems to be the most neglected; and, shame to say, it is yet under consideration, if music shall be admitted as a science. It would be out of place to treat on the science of music in this book; but as I have already touched the subject, I will finish by making a few remarks.

The best performers, who have natural talent for music, does not require the science of the same; but for composers, it is of great importance, and for musical instrument makers, it is indispensable. One reason why musical taste is so little cultivated, is for the want of good instruments, and the whole world is filled up with musical instruments of inferior quality, for the want of science in their manufacture. Many of them make perfect cat music, when it is unreasonable to expect of the performer to cultivate musical taste. I have often wished to put down such musicians to the grade of organ-blowers, but reflecting on the subject, I find myself greatly in error—it is the instrument maker who is to be blamed, and not the performer.

The only wind instrument constructed on purely scientific principles, and which has attained perfection, is the *Boehm Flute*, which, I believe, is the best example to show what science can do in music. Mr. Boehm himself is an excellent performer on the Flute, he is a good composer, a skillful mechanic, with inventive ingenuity, and he is a scientific man. Every one of these faculties are brought to bear on the construction of his Flute, and it could not have been perfected without one of them.

In the year 1858, I had the pleasure of Mr. Boehm's acquaintance, in Munich, when he was constructing a G Flute, which is $2\frac{1}{2}$ notes below our ordinary C Flute. Mr. Boehm remarked, that the finest musical ear cannot detect a difference in a tone so delicate as he can establish it by calculation and measurement. I examined very closely the details of the holes and keys on the G Flute, which was made of silver, and found that no allowance was made for retoning it by trial and error, but it was made right from the first design.

The finest musical wind-instrument in existence, I believe, is the Bassoon, (Fagott); but for the want of science in constructing it, it is now very much declining. If a Bassoon was constructed under similar treatment as that of Boehm's Flute, I believe it would produce a sound approaching the finest human voice. The tones on the Bassoon sound very much like one singing in the nose, which is a clear indication that the musical vibration is jammed up at the point where it is broken off.

Abreviation of the Tonal Units.

M = *Meter*, unit for length.
G = *Gall*, unit for capacity.
T = *Tim*, unit for time and the circle.
P = *Pon*, unit for weight.
H = *Horse-power*, unit for work.
D = *Dollar*, unit for money.
Tp = *Temp*, unit for temperature.

The abridgment of the units to be noted by capital letters, and the multiplication and division of the same as an exponent by a small letter placed before or after the unit, thus, M^t = Meterton, ^tM = Tonmeter, G^s = Gallsan, T^s = Timsan, P^m = Ponmills, &c., &c.

The sound of the new names will of course be strange to the ear, before its corresponding character and value is known by heart, but as before stated, the object aimed at has been to select such sounds as would become best suited for all languages, and at the same time be simple and expressive.

EXAMPLE 11. When 3·꼰9P of butter cost 4·3SD, how much will 1꼰·2꼰P of the same cost?

$$3·ꂝ9 : 1ꂝ·2ꂝ = 4·3S : X$$
$$X = \frac{1ꂝ·2ꂝ \times 4·3S}{3·ꂝ9} = 1ꂝ·36 \text{ dollars.}$$

```
 1ꂝ·2ꂝ              3·ꂝ9 | 72·9910 | 1ꂝ·36
 4·3S                      3ꂝ9 ...
 ─────                     ──────
 ꂝꂝ70                     3609 ..
 51S9                      350ꂝ ..
 6ꂝꂝS                     ─────
 ─────                      ꂝꂝ1 .
 72·9910                    ꂝ5ꂝ .
                           ─────
                           1S30
                           16ꂝꂝ
                           ─────
                           174
```

The answer is 1ꂝ dollars, 3 shillings and 6 cents.

EXAMPLE 12. What will be the interest on 3ꂝꂝS·65D in 4 years and 5 months, at 9 per sant per annum? (9 per sant is about 6 per cent. decimal.)
Interest = 3ꂝꂝS·65 × 0·09 × 4·5 = 953·S2 dollars.

EXAMPLE 13. A yearly payment or annuity of 3ꂝꂝ·65D is standing for 15 years and ꂝ months,—what will it amount to in that time at ꂝ per sant interest?

Amount = 3ꂝꂝ·65 × 15·ꂝ [1 + $\frac{0·0ꂝ}{2}$ (15·ꂝ + 1)] = SS4ꂝ·ꂝꂝ dollars.

```
  3ꂝꂝ·65            1ꂝ·ꂝ              5473·127
    15·ꂝ            0·005S             1·5ꂝꂝS
  ───────           ──────             ─────────
  244357             ꂝ5S               2935S53S
  135SSꂝ              Sꂝ7              3ꂝ564ꂝꂝ4 .
  3ꂝꂝ65             ─────              445ꂝ7ꂝꂝꂝ ..
  ───────           0·5ꂝꂝS             2ꂝ50ꂝ95ꂝ ...
  5473·127                             5473127 ....
                                       ─────────────
                                       SS4ꂝ·ꂝꂝ96178
```

EXAMPLE 14. When one gall. of fresh water weighs one pon, how much will 5·8̆ằ cubic meters of cast-iron weigh, when the specific gravity of the iron is 7·212?

Weight 7·212 × 5·8ằ = 25·8463ʊ Pons, the answer.

A similar example by the old system will be very complicated. Even in the French metrical system there is more confusion in pointing off the decimals.

EXAMPLE 15. A locomotive running 39·ằᵐM per *Tim*, leaves London at 5·84T; another locomotive on the same track makes 6ằ·ʊᵐM per *Tim*, leaves London at 6·2ɩT. At how many Millmeters from London, and at what time will the faster locomotive reach the other?

The time of the fast locomotive will be,

$$T = \frac{39\cdot\text{ằ}\ (6\cdot2\ell - 5\cdot84)}{6\cdot\text{ằ}x - 39\cdot\text{ằ}} = 0\cdot7 \text{ tims, the answer,}$$

and 6·2ʊ + 0·7 = 6·5ʊ tims, the time when the fast locomotive reaches the other. Distance from London will be 6ằ·ʊ × 0·7 = 30·74 Millmeters.

Examples in Navigation, Comparing the Old and Tonal Systems.

EXAMPLE 16. In the year 1861 the sun's declination at Greenwich mean noon is:

	Old System.		Tonal System.
March 13,	2° 48′ 36·9″	=	0·1ff9 tims.
March 14,	2° 27′ 57·3″	=	0·1ɩ78 "
Difference,	23′ 39·6″	=	0·0478 "

Required the true declination at mean noon, on the 13th of March, 1861, in longitude west from Greenwich, 156° 40′ 23″ = 6·f70ɔ tims = 0·6f70ɔ days.

54

Old System.

```
     0·4353              156°                360
       23 Mult.           60 Mult.            60 Mult.
    ───────             ──────             ──────
    13059               9360               21600 Min.
     8706                 42 Add.             60 Mult.
    ───────             ──────             ──────
    10·0119 Min.        9402 Min.         1296000 Sec.
                          60 Mult.
     0·0119            ──────
        60 Mult.       564120
    ───────               23 Add.
     0·7140            ──────
Add 39·6 Seconds.      564143 Seconds.
    ───────
    40·3140
       40 314            564143·0000  | 1296000
        0·4353           5184000 ...  |─────────
    ───────              ───────...   |  0·4353
       120942            4574300 ..
       201570.           3888000 ..
       120942..          ───────
       161256...         6863000 .
    ───────              6480000 .
    17·5486842 Seconds.  ───────
                         3830000
                         3888000
```

from which the correction will be 10' 17·548".

Tonal System.

```
Diff. longitude        0·6f705 days.
Diff. declination      0·0478 tims.
                      ─────────
                       5?0275
                       36f13f
                       188f24
                      ─────────
         Correction    0·01f934365 tims.
```

The required declination will be,

$$2° 48' 36.9'' = 0.1𝔙9 \text{ tims.}$$
Correction, add $10' 17.5'' = 0.01𝔙9$ "

True decli. $2° 58' 54.4'' = 0.21𝔙4$ tims, the answer.

The old system required seven multiplications, one division, four additions, and employed in the calculation about 215 figures, while the *tonal system* required only one multiplication, and employed only 39 figures for the same object, namely, to find the correction.

The old system required a great deal more knowledge of how to manage the many different operations and figures, and in consequence subject to more errors, while the *tonal* system is simple, clear, and natural to the mind. In working the time and lunar observations, the difference will be still more, on account of angles being expressed both in time and degrees.

Although our decimal arithmetic is based on 10, the ordinary mind has found it more simple to divide their units into 8, 12, 16, 20, &c., &c., parts, but as the science in arithmetic advanced, the educated mind found that the units could also be divided into the unsuitable number 10 parts, and so the complication was carried into weight and measure.

Had the *tonal system* been adopted instead of the decimal arithmetic, it would surely be considered ridiculous to divide units into 10 parts.

In the measurement of machinery there are very few dimensions that come up to the length of the French *meter*, then most of the measurement must be expressed in decimal fractions, often with several cyphers before the figures as is readily seen on French drawings. A French millimeter $0.001^{m.} = 0.039$ inches.

A *tonal* santimeter $0.01^{m.} = 0.0229$ inches. Here the French system require one cypher more in expressing a quantity 58 per cent. greater than the tonal expression, and the difference will be much more in squares and cubes.

In calculating the weight and cubic contents of machinery we have often very small dimensions to deal with, then a cubic centimeter will be $0.000001^{c.\ m.}$ which by the *tonal* system will be only $0.001^{c.\ m.}$ or three cyphers less for the same dimension. Decimal fractions of this kind make more or less confusion by the many cyphers, in squaring or cubing numbers, and more so in extracting roots.

The French metrical system is, however, not uniform throughout, by which it is self-evident that the meter is too large. The unit for capacity one *litre* $= 0.001$ cubic meter, and the unit *gramme* $= 0\ 000001$ cubic meter of distilled water in weight.

In the tonal system there are no such irregularities. One cubic meter is the unit for capacity, and the weight of one cubic meter of distilled water is the unit for weight.

It is self-evident that the French metrical system is not well suited in practice, for although it has been enforced by law in France for over twenty-three years, yet, expressions of the old system are frequently used in the shop and the market; and oftentimes particularly in the interior, bargains are made in the old system and the bills made out in the new system. In the Paris market, it is most invariably heard that an article cost so many *sous* per *livre*, and it is very natural that it remains so, because the decimal system is too troublesome. It is, of course, easier to count twenty sous on the franc instead of one hundred centimes.

In dividing things, it often happens that 10 parts will be too fine divisions, or the delicacy of the work may not require all the 10 parts; it is then suggested to take only every other part, but in so doing it does not fall in with half the base 5, for which it may be necessary to reject either five or none of the divisions, which is a great inconvenience in practice. On thermometer scales, we often find every other degree marked, which does not come in with the fives.

In the *tonal* system, every other or every four parts will fall in with the halfs and quarters, as well as with the base.

THE COUNTING MACHINE.

The counting machine, fig. 6, consists of a square frame, in which are inserted ten brass wires or lines $a\ b\ c - k$, eight of which have each 10 *tonal* balls, moveable from one side to the other. The void line c, is to denote *tonal fractions*, that is, the balls on the line a, to denote cents, metersans, timsans, &c., &c., and the balls b for shillings, metertons, galltons, &c., &c., d the unit, e 10, f 100, g 1000, i 10000 and k 100000. The line h, separates the *mill* and *bong*, to make the reading more clear.

Before the operation of counting is commenced, all the balls are to be on the left side, and to be moved towards the right as the counting requires. The instrument is to be used principally for addition and subtraction, but multiplication and division can also be performed on it, with some assistance of mental calculation.

Fig. 6.

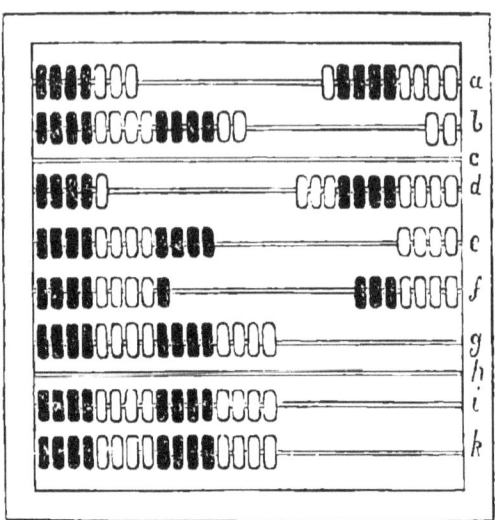

The operation on the counter is similar to that on the ancient *abacus*.

The number noted on the counter is 748·25, which according to whatever unit it means, may be 748 dollars, 2 shillings and 5 cents, of which 700 is on *f*, 40 on *e*, and 8 on *d*. Suppose the sum of 1345·42 dollars is to be added to 748·25D. Move one ball to *g*, which denotes 1000, 3 balls towards *f*, 4 towards *e*, 5 towards *d*, 4 towards *b*, and 2 towards *a*; and the sum will be found on the left side 1950·68 dollars. The operation is generally commenced on the top line *a*, and when there are not balls enough on the line, subtract the complement of 10 and add one on the next line below. This will be understood by the following example:

Add 39 cents to the 25 in the noted sum. Say 9 from 10 is 6, move 6 from *a*, and there will be 3 left, move one towards *b*, move further 3 balls towards *b*, and the result will be 63 cents, or 6 shillings and 3 cents, the sum of 25 + 39. Add further ℓ8 dollars to 74ℓ. Say 8 from 10 is 8, move 8 balls from *d* and 1 to *e*; say ℓ from 10 is 4, move 4 balls from *e*, and 1 to *f*, and the sum will be 74ℓ + ℓ8 = 813 dollars.

In this manner, addition can be continued on the counter up to 1,0000,0000 *tonal*, or 53,736,276,736 decimal.

By the aid of this instrument, the *tonal system* would be easily acquired, because it turns the mind from the old base, which is of the greatest importance; besides, the counter would be a most valuable instrument for adding columns of figures.

THE RUSSIAN STCHOTY, (*Fig. 7.*)

This instrument is in common use in Russia. Every counting-house, office, store, or shop, of whatever description, and every family has a *stchoty*; in fact, it is as common in Russia as a spoon or knife. In the steps of southern Russia, where a house is rarely to be found, and where it is difficult to find anything to eat or drink, a *stchoty* is always to be found, even among the Kalmuks; and it is surprising to see with what readiness and correctness the Russians use this instrument for their calculations. They can multiply and divide with great facility on it.

The operation on the *stchoty* is the same as that described for the counter. The line *a* denotes quarters of copeks, *b* single copeks, *c* 10 copeks, *d* 25 copeks or $\frac{1}{4}$ of a ruble for each ball, *e* rubles, *f* 10 rubles, *g* 100 rubles, *h* 1,000 rubles, and *i* 10,000 rubles. The sum noted on the *stchoty* is 743·45 rubles or 743 rubles and 45 copeks.

The line *d* is not absolutely necessary, as the copeks

Letter from the International Association.

10 FARRAR'S BUILDING, TEMPLE,
LONDON, 31st October, 1859.

SIR:—

Your esteemed letter of the 1st of June, has been received, together with a copy of your description of the *Calculator*, which is evidently a most ingenious and useful instrument, and the manuscript account of your new system of arithmetic and measures, weights and coins.

We think ourselves much honored by the confidence which you have manifested towards us, and are of opinion that we shall best testify our high sense of your ability and intelligence, and your zeal for the improvement of mankind by the most free and sincere expressions of our sentiments upon your project.

The project has evidently the great merit, which, as far as we know, belongs to no other hitherto invented, except the metrical system, that is a *uniform system* founded upon one simple principle, which is consistently applied throughout to the attainment of its professed design. We are, nevertheless, sorry that we cannot give it our support, having by the very constitution of our society, and from its first foundation, adopted the number 10 as the basis for such a system. On reviewing the grounds of our original determination we see no reason to depart from it.

When the metrical system was invented by the careful deliberation of the first mathematicians of the age, they studied the question by the arithmetical scale, and especially to take 12 as a basis, because that number seemed to have in some respects a claim to be taken in preference to 10. After a full examination of the question they decided that it was necessary to retain 10 as the basis in arithmetic, and to adopt it universally for measures, weights and coins.

Your system would be far more difficult to learn than the other. When learnt, it would require a smaller number of figures in each operation, and might therefore present some facilities for making calculations in writing, but it would be very burdensome to the memory so as to be unsuitable for mental arithmetic, and consequently for all the smaller dealings of the shop and the market, and for those minute calculations which in all arts, trades and manufactures often requires to be performed with the greatest possible rapidity. There is a limit to the powers of the human mind, and it appears probable that, except in extraordinary cases a system founded on 16 as a basis, would be found to exceed the natural capacity of man for the use of numbers.

You object to the meter, as "much too long to be convenient to the artizan, and you therefore choose for your unit a length which is about the seventh part of a meter. The proper aim in determining upon a unit of length is to find one adopted, as far as possible, to all uses without exceptions, and the general consent of mankind seems to point to the conclusion that a length approaching to the meter best corresponds with this intention. For specific purposes the meter is divided or multiplied either by 2 or by 5, and thus you may obtain any measure you please including your own unit which is very nearly equal to 15 centimeters.

We could show you, if we had the pleasure to see you here, numerous decimal divisions of the meter such as the measures 5, 10, 20, 25, 30, 40, and 50 centimeters. We have these graduated down to half milimeters, made of a great variety of substances, and with considerable difference of form, either solid and in one piece, or made to fold with hinges, or to be wound on roulettes in cases so as to be carried in the pocket with the greatest ease imaginable.

Also some are made with slides. In short the meter is proved by experience, an experience which is extending every day over wider and wider area of the earth's surface, to be adopted for the artizan as well as for every other occupation. You express a preference for the English foot, but if the foot has advantages, you may take the foot of Hesse Darmstadt which is exactly the fourth part of the meter. It is considerably nearer than the English foot to your own proposed unit, and in itself is unquestionably as convenient for the artizan as any other foot.

In conclusion, we beg to present you with the principal publications which have been issued by our branch of the International Association.

We would especially direct your attention to our treatise on the best unit of length. In section VI. you will find a discussion of the question respecting its adaptation to binary division, and in section VIII. it is maintained in opposition to your views, that the meter may be employed with the greatest possible advantage in the mechanical arts. You will allow us sir, to indulge the hope that further examination of the subject may induce you to coincide in the opinion which we endeavor to defend and which is gaining ground, as we understand, in Russia, as well as in the civilized countries.

With much respect we subscribe ourselves on behalf of the British Branch of the International Association.
Your obedient servants,

(Signed,) JAMES YATES, F. R. S., *Vice President.*
LEONE LEVI, *Resident Secretary.*

MR. JOHN W. NYSTROM,
Rostof on the Don.

P. S.—Further observations on the same topic are published in Lord Overstone's questions with the answers, London Folio, 1857, 176, 179. Your letter and pamphlet were shown at a meeting of the International Association, at Bradford on the 10th, 11th, and 12th ultimo, and we were desired to convey to you their thanks for the valuable suggestion you have offered.

LEONF LEVI, *Res. Sec.*

LUDINOVA, IN THE GOVERNMENT OF
KALUGA, RUSSIA, Nov. 26, 1859.

To the International Association for obtaining a uniform Decimal System of Measures, Weights, and Coins, No. 10 Farrar's Buildings, Temple, London:

GENTLEMEN:

I have had the honor to receive your favor of the 31st of Oct'r last, for which I beg to return my most sincere thanks. Feel very much gratified indeed that your honorable body considered my suggestions worthy of notice.

I hope the International Association will bear with me for making some remarks on your letter, by which

I have no other object, but to discuss the connection of practical and scientific principles, and sincerely beg you to pardon my straightforward expressions.

It seems to me that the real substance of my project for the *tonal system* of arithmetic, weight, measure, and coins, is not well conceived or appreciated by the International Association, because there are statements in the letters which are not exactly in accordance with the fact, likely originated from the difficulty in conceiving a new arithmetical system with a new base, when the base 10 is impressed on the mind. You state that my "System would be far'more difficult to learn "than the other." Such is not the case; it is *easier* learned, and although the multiplication table of the single figures is about two and a half the extent of that of the old system, it is much easier acquired.

The only difficulty is, to turn the mind from the old base: for instance, $8 \times 8 = 40$, will appear very curious to a stranger, who is perfectly sure that $8 \times 8 = 64$, but when he knows that 16 is the base for the system, and 8 is half of the base, he will easily conceive that half of 8 is 4, and $8 \times 8 = 40$.

"When learned, it would require a smaller number "of figures in each operation, and might therefore pre- "sent some facilities for making calculations in writing." This is of little or no importance in the sense you apply it: it makes very little difference if one or two figures, more or less, are written down on paper, which is a mere mechanical operation compared with having the figures clearly located on the mind.

Any measure under 8 feet can, by the tonal system, be expressed to a nicety of a fraction of a millimetre, or less than 32nds of an inch, with only three figures, which, in ordinary cases, would require at least five figures on the French metre; and a measure of up to 122 feet is expressed to the same nicety by only four Tonal figures. The utility of the tonal system is not limited by the small number of figures expressing a delicate measure, but on account of the figures at the same time impressing the mind of the natural fractions, quarters, eighths, and sixteenths, the principal utility lays in the clearness of the expression. Sixteenths expressed by decimals will require four times the number of figures, which will carry the mind 1000 times further than the tonal system. In astronomical and nautical calculations, there will be required a less number of figures, on account of dispensing with a great number of tables.

"But it would be very troublesome to the memory, "so as to be unsuitable for mental arithmetic, and "consequently for all the smaller dealings of the shop "and market, and for those minute calculations which "in all arts, trades, and manufactures often require to "be performed with the greatest possible rapidity."

This is not right, and it proves that the utility of the tonal system is not within your comprehension. For mental calculations, the shop and the market, it is best suited, and the very reason why I have proposed it. For mental calculation in addition and substraction, the mind need not be carried further than the base 10, and in multiplication and division only to 100. I would not take the trouble to invent or propose a new system of arithmetic for "the greatest "mathematician of the age," to whom it makes little or no difference, if the base is a prime number, for his *ps*, *qs*, and *ƒs* are applicable to any system whatever; neither would I think it worth the while to alter the system of arithmetic for the small portion of the public who have to do with quantities only by pen and ink, for whom it is very easy to find a suitable system, and the decimal system with 10 or 100 to the base will answer that purpose fully; but it is not so in the shop and in the market, where the natural fractions and aliquot numbers are wanted, as well for mental calculations as for the mechanical divisions and proportions of materials.

"There is a limit to the power of the human mind, "and it appears probable that, except in extraordinary "cases, a system founded on 16 as a basis would be "found to exceed the natural capacity of men for the "use of numbers." The decimal system is the worst that ever could be selected of the even numbers in the neighborhood of 10, eight or twelve would require less capacity of mind. It is easily found in practice that when the base of a measurement is 12 or 16, it is easier managed in the mind, even with the present system, and it would become so much easier if the Arithmetic had the same base.

Had I proposed 11, 13, 14, 15, or 17, to the base, the system would become more difficult in a quadruple or triple proportion to the number of new digits added, but 16 is quite an exception to that supposition; 9 as a base would be a great deal more difficult than 10.

I will here give you some numbers placed in order as they become more difficult as a base for Arithmetic, namely, 8, 16, 12, 10, 14, 9, 11, 15, 13. It can very readily be conceived that the present arithmetical base is too small, because, referring to the decimal system in practice, it will be found that it is generally impressed on the people's mind that the unit is divided into 100 parts, by which quantities are generally written and expressed, for, although it is at the same time divided into 10 parts, it is seldom used so. For instance, in France, it is never said or written that a measure is so many decimetres long, but it is expressed in so many centimetres.

In America, it is never said or written that an article cost so many dimes, but it is expressed in so many cents. We have also in America, from the Spanish money, the dollar, divided into 8 and 16 parts, which are mostly expressed by separate names, and in reality found to be a more suitable division for the market. Suppose an article to cost 38 cents, and it is paid with a dollar. Now, the seller must carry his mind to 100, and then back to somewhere about 70; here he will be confused about the eight, and not sure if it will be correct to subtract the 8 from 70; but finally he finds out that it is 62 cents to be returned on the dollar; the buyer—most frequently not so smart in counting as the seller—will perhaps say that there should be 72 cents change. This example I have given from actual and frequent observation in practice. Now, suppose a similar example with English money: an article cost 38 pence; it will be observed that 38 pence is not noted, but it is said or written three shillings and two pence. Suppose the buyer to pay for the article with a crown, which is five shillings. The seller will very likely reply, "Have you two pence, and I will give you two shillings?" or he may give the buyer 1s. 10d., and so the affair will end with perfect understanding; the mind was not carried above the base 12, while in the

American case the mind was carried more than eight times further, namely, to 100. The decimal system is therefore very troublesome for mental calculation, and frequently approaches the "*limit to the power of the human mind*," which would be rarely the case with the tonal system. It will not be denied that halfs, quarters, eighths, and sixteenths are the most natural fractions for the artizan, shop, and market, and they are frequently expressed by decimal fractions; but if 0·125 is shown to the majority of the people, there will be comparatively none who understand the true meaning of it; and if it is told to them that 0·125 means $\frac{1}{8}$, it will be necessary to explain that the whole is divided into 1000 parts, and 125 of the parts is $\frac{1}{8}$ of the whole. The people will then surely reply that this is a roundabout way of doing things, and that they are not willing to cut their things up into 1000 parts in order to get it into eighths. I am inclined to believe that among the best arithmeticians, including the International Association, there are few, if any, who clearly comprehend that 125 is $\frac{1}{8}$ of 1000, but it is well known to be so by practice in calculation. It is easy to comprehend that 25 is $\frac{1}{4}$ of 100, from which it can be conceived that $\frac{1}{2} \times 25 = 12·5$, and by that way it may be impressed on the mind that 125 is $\frac{1}{8}$ of 1000. If 12 was the arithmetical base, it is easily conceived that $\frac{1}{8} = 0·16$, but with the tonal system it is most easy to comprehend that $\frac{1}{8} = 0·2$. Therefore the decimal system is very complicated and difficult, as well for mental calculation as for the artizan's ordinary application of numbers and measurement. A high number of several digits must be managed in the mind, in order to comprehend a small one of only one digit.

In music, the tonal system is in full operation; the notes are divided, as regards time, into halves, quarters, eighths, &c., &c.

Fig. 8.

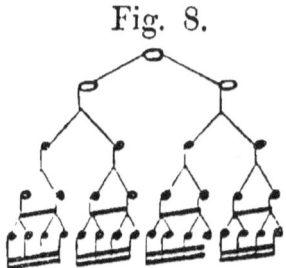

A bar of music is generally expressed by quarters or eighths, and a burden has generally 8 or 16 bars.

Now, suppose that a musician is requested to divide his notes, bars and burdens into fifths or tenths, according to the decimal system,

Fig. 5.

thus 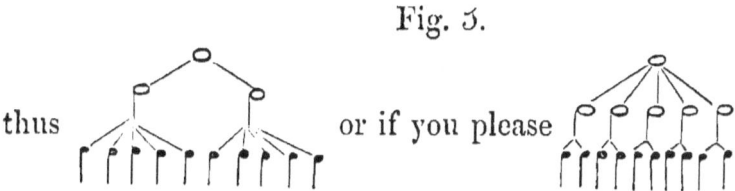 or if you please

then ask the musician to play a decimal piece of music, and it will sound very much like the decimal system introduced in the shops and markets. This is the best comparison I can give between the tonal and decimal systems, because, if the world was fortunate enough to be in possession of the tonal system, and knew nothing about it, but requested to turn the mind towards the decimal system, it would be much more awkward to mankind than for the present musician to do so.

"You object to the metre as much too long to be convenient to the artizan,"—Yes; and the International Association has given me a further proof of the fact, which I beg to explain hereafter,—"and you "therefore choose for your unit a length which is about "the eighth part of a metre." It is stated in the manuscript that the length of a step of a man, or about two feet, appears to be a suitable unit, and when I divided the circumference of the earth with 16^7 it was my greatest wish to arrive at a unit of about 20 to 28 inches, but as the length of the assumed main standard was not under my control, I was obliged to be contented with the last quotient 5·865 inches. It was, however, my intention to propose to divide the quadrant of the earth with 16^6, which will give a unit of about $23\frac{1}{2}$ inches, but in order to follow a uniform and unbroken system of division throughout all kinds of measurement, I concluded to maintain the first quotient of 5·865 inches as the unit for length.

"The proper aim in determining upon a unit of "length is to find one adopted, as far as possible, to all "uses without exception."

That is just the very object of my aim, and it is the inconvenience and defalcation of the decimal and metrical system, that has called on me to propose something better. It seems to me that in making such statements, it would have been well to give some reason and example where the tonal system is inapplicable.

The metrical system is inapplicable in navigation, because it does not agree with the degrees and minutes of the great circle of the earth, which also makes some inconvenience in geographical survey. The decimal system cannot well be adapted for the division of the circle and the time, nor can it be adapted in music, which forms the most natural conception of division. "The general consent of mankind seems to point to "the conclusion that the length approaching to the "metre best corresponds with this intention." This is not correct. If the table of foreign measures of length is examined, it will be found that the whole world points towards the English foot, that the French metre stands alone the longest measure, and that it is only the Persian arshine which attempts to approach it. Nations point towards the French uniform decimal system, merely because it is, as far as our present arithmetical system permits, in itself the most complete for calculation, but if the French had adopted a shorter metre, I believe the system would have been picked up much sooner by other nations.

" For special purposes the metre is divided or multi-
" plied either by two or by five, and thus you may
" obtain any measure you please, including your own
" unit, which is very nearly equal to 15 centimetres."
That I do not understand.

" We could show you, if we had the pleasure to see
" you here, numerous decimal divisions of the metre,
" such as the measures of 5, 10, 20, 25, 30, 40, and 50
" centimetres." This is a proof that the metre is too long, and very likely some practical mechanic or engineer has made the same remark, for which the metre is cut up into pieces, in order to show that it can be made shorter. Let us examine the pieces one by one. A measure of only 5 centimetres is of little importance to the artizan, besides 5 is a prime number, which makes the whole decimal system objectionable. A length of 10 centimetres is very convenient for minute measurement, but too small for general use. Twenty centimetres is a good measure within itself, may be conveniently used in the drawing-room and for

measures not exceeding its length, but for more than 20 centimetres, it will be accompanied with an objectionable mental calculation. It is contained in the unit five times, which is a perplex number for the artizan, because when the fifth part is laid down he may be uncertain whether he has laid down four or five 20 centimetres, and when he looks back on the measured part, he cannot well conceive the correctness without going over it once more, which would not be the case if the measure was contained four times in the unit, where the halfs and halves certify the correctness. A measure of 25 centimetres has the only advantage of being contained 4 times in the metre, but within itself an unsuitable measure; in measuring off a distance between 25 and 100 centimetres, it is accompanied with a troublesome mental calculation.

Twenty-five centimetres are rather long for the pocket; it must be folded, but into how many parts? if folded into two parts, there will be $12\frac{1}{2}$ in each. Thirty and forty centimetres are not evenly contained in the unit, and will, in practice, be accompanied with troublesome mental calculation.

Fifty centimetres chopped up into four or five parts, has its evident disadvantages. If these different measures are introduced into the market, people will become accustomed, one to a 20 centimetre, another to a 25 centimetre, and some select 30, 40, or 50 centimetres; then when one gets hold of a strange centimetre, he is apt to make a mistake in his measurement and calculation.

" We have these graduated down to half millimetres, " made of a great variety of substances, and with

"considerable difference of form, either solid and in
"piece, or made to fold with hinges or to be wound on
"rollers in cases, so as to be carried in the pocket with
"the greatest possible ease imaginable."

I am perfectly convinced that the metre can be made convenient for the pocket, but I say that it is not convenient for measurement and mental calculation, and I am sure that it requires a great many ingenious contrivances to put the metre into a suitable shape, but among all your varieties of metres and centimetres, have you a single sample which can practically be considered so good a measure as the English two-foot rule? You will allow me to doubt it. I have also in my possession a few varieties of the metre, but none which I consider a proper measurement, and I have never seen a good metre even in France, although I have made great efforts to procure the best possible. Those in my possession are all made to fold into 10 parts, made of ivory, brass, fish-bone, wood, and one a tape to roll in a case, but they are all toys. In Marseilles once I bought a metre of the ordinary form, made of ivory, to fold into 10 parts, went home to my hotel and tested the metre on my standard rule, and found it to be $1\frac{1}{2}$ millimetres too short. I returned immediately to the instrument maker, Mr. Santi, No. 6 Ferreol Street, stated the fact, which was soon testified on a standard metre, and I was offered to select a correct one, which made me try a great many metres one by one, and did not find two of the same length. I then suspected the great many joints, tried several by pushing and drawing, when I found a little motion in some of them, tried again two metres, the shorter one I stretched a little, when it became the same length as the other. I selected one metre by the standard in the store, which I have now on my table; it has grown two millimetres longer when I stretch it out, and when I push all the joints in an opposite direction the metre will be $\frac{1}{2}$ millimetre too short.

I do not blame the workmanship of the metre in question, because it is made as good as it can be, and it is equally good as those I selected in Paris, where I found similar metres to those in Marseilles; but I object to the principle of the instrument, because it is in every shape inconvenient in practice. It is very inconvenient to lay out work by the ordinary pocket metre; for instance, the metre must be kept and adjusted by the left hand at a, and stretched by the

right hand at b, it is then required a third hand to straighten the decimetres between a and b, because the work is oftentimes such that if the hand is taken from b, that end of the meter will fall down, and disturb the adjustment at a.

In practice it often happens that it is inconvenient to get at one of the points between which a measure is wanted, a two foot rule is then stuck over to the furthest or otherwise inaccessible point; and the measure read at the nearest point; in a great many such cases of daily occurrence, it would be impossible to employ direct a ten-folded French metre, for which the two hands are required, one at each point. Suppose the outside diameter of a cylinder is to be measured, it is generally taken in a pair of callipers, then by the

English mode the callipers are kept in the left, and the rule in the right hand, while the diameter is read; now by the French measurement two hands must be employed to keep the metre while a second person must be employed to keep the callipers. You will now surely remark that "the metre can be made to "fold with hinges into four parts—similar to the "English rule, and used with the same advantage"—to which I beg to reply that the metre in such a form will be rather clumsy for the pocket, and for the artizan, and on account of its great length it will not have the firmness of an English rule. Two and a half decimetre or the odd number of 25 centimetre in each part is an indication that there is something wrong about it. Half a metre folded into two or four parts is a broken up half thing—I say broken up, because two parts will contain each an odd number of divisions 25, and four parts, will contain each $12\frac{1}{2}$ centimetres.

Another measure generally employed as a standard by architects, city surveyors, in machine shops, &c., &c., about 8 to 12 feet long, and very likely in the office of the London City Survey will be found standards of 10 feet, which is a very convenient measure for a great many out-door works; a measure of that kind will be about three to four metres, which are very inconvenient numbers—accompanied with extra calculations in laying out a long measure for which a tape or a chain is not correct enough. If such a measure is made five metres, it will be rather long and inconvenient, and accompanied with a mental calculation which by the prime number 5 gives an odd number at every other operation.

A measure of 10 metres cannot well be employed in the streets, except in the form of a tape line or a chain, but for such form 10 metres is too short. A tape line or a chain ought to be about 50 to 100 feet long. Further you state that, "in short, the metre is "proved by experience, an experience which is extend-"ing every day over wider and wider area of the "earth's surface, to be adopted for the artizan as well "as for every other occupation." This is saying much.

Has experience ever had anything to do with the length of the metre from its very first origin? It was according to your statement "*invented by the first* "*mathematicians of the age*," after which it was intruded on the French artizan by law, from which experience in using it, was necessarily attained. The mathematician had the measure of a quadrant given to him in figures, which he found was easiest to divide by 10s in order to arrive at a small number, but had the mathematician been set into practice to divide a quadrant or a straight line by a pair of compasses, he might have discovered that the most easy, and the most correct divisions are attained by dividing it into half and halves, which would have given a quotient of about $23\frac{1}{2}$ inches as a metre. The length of the metre has nothing whatever to do with the utility of the French uniform decimal system of weight, measure and coin. Had a shorter metre been adopted, and such a cubic metre of distilled water called a killogram, the same advantage would have been attained. The principal difficulty in introducing the decimal system, and the general discord of weight and measure throughout the world, is caused by the unsuitable base in the arithmetical system.

It would indeed be a great service to mankind, if some leaders of the scientific world would deviate a little from their determination to maintain and propagate a system so ill-suited for the purpose for which it has so long toiled.

"You express a preference for the English foot, "but if the foot has advantages, you may take the foot "of Hesse Darmstadt which is exactly the fourth part "of a metre." The Darmstadt foot would do me precisely the same service as any other of the nearly 35 different foots employed in different parts of the world, I do not care how many times it is contained in a metre, because that has nothing to do with the subject in question. The object aimed at is to invent, propose and introduce to the world a system of calculation, weight, measure and coins, which would without exception fulfil all requirements of mankind, and when such is attained, it makes no difference how many times the Hesse Darmstadt foot goes in a metre.

I thank you most sincerely for your kindness in offering to me the principal publications issued by your branch of the International Association and hope soon to receive them, and I shall read them with the greatest interest. I am well convinced that you have furnished plenty of good materials in favor of the decimal and metrical system, which are current among a great many of your readers. You can give many instances where the metre can be conveniently employed; you can give examples that when so many tons, cwts. and pounds is multiplied by so many £ sterling shillings and pence, and divided by so many fathoms, feet and inches, will be a long and complicated calculation, compared with the measures at once expressed by decimals; besides the metrical and decimal system being adopted and in successful operation, by one of the first empires in the world, is indeed a great temptation.

According to your statement, I expect to find in "Section VIII., it is maintained in opposition to my "views that the metre may be employed with the "greatest possible advantage in the mechanical arts." Such is easily maintained in writing, but go to practice, and give an English mechanic a French metre of the ordinary ten-folded form, and ask him to measure a distance of about 20 inches; the mechanic will then fumble about in straightening the decimetres, and if there is no support between the two points, he will hang the metre in a catenary form, as *he is* not accustomed to employ two hands for such a small measure, —he will then very likely tell you that the measure between the points is 50 and some small marks which he cannot read. Now give an English 2 foot rule to a French mechanic, to measure a similar distance, and he will tell you immediately without hesitation that it is 20 and $\frac{3}{8}$.

The inches being divided into halves, quarters and eighths, makes the reading so clear, that the very first glance impresses the mind of the correct measure.

A captain sailing along the sea cost in a dark night, requires to be if possible always in sight of a light, in order to be sure of his position, and safety of his ship; such is the case with the mind sailing along a graduated measurement. On the English rule, fig. v, there are big lights at short intervals, and beacons and buoys between them, while in the wilderness on the French metre, fig. v, you encounter sometimes a little light high up in the arithmetic atmosphere, looking very

Fig. ß.

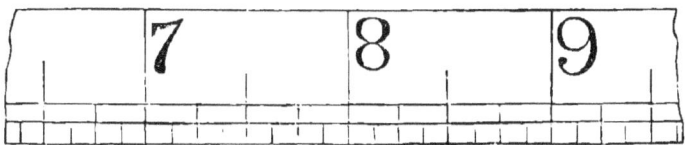

much like the old street oil lamps before lighting gas was invented, and between them you encounter a number of things one like the other, by which you are not

Fig. ѵ.

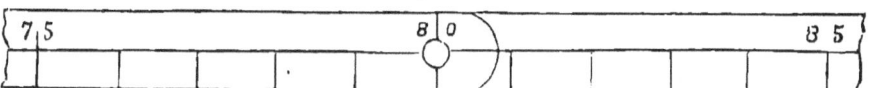

sure whether you are here or there. The ordinary English rule such as made by Mr. Elliott, London, or Field and Son, Birmingham, will stand and measure as long as twenty French pocket metres, and it will measure the last piece as correct as the first one, which is not the case with a ten-folded metre stretched a few times.

On mathematical instruments in general, the decimal division is very troublesome, compared with the natural divisions, for instance, in verniers, fig. ε, this makes a clear reading, and divides the inch into 256 parts, while the decimal system, fig. z, is more difficult and divides the inch into only 100 parts.

Fig. ε.   Fig. z.

I regret very much to say that the closer I examine the subject, the more I am inclined to oppose the

French metre, as well as the decimal system, which is in reality the most unnatural system of division, which could reasonably be selected.

I am sure that in a thorough practical examination the metre will stand a poor chance, and I shall be much mortified if the law intrudes upon me such inconvenient measurement for my mechanical works.

Weight.

The decimal system is equally inconvenient for weight as for all other measurements, the unit being divided into 10 parts, for which are required five different weights in weighing all the ordinal parts namely 1, 2, 3, 5, and 10, or a weight of 4 may be substituted for the 3, but it is at any rate an odd and dreary composition of weights.

$$1 = 1 \text{ weight.}$$
$$2 = 2 \text{ weights.}$$
$$3 = 3 \text{ "}$$
$$3 + 1 = 4 \text{ "}$$
$$5 = 5 \text{ "}$$
$$5 + 1 = 6 \text{ "}$$
$$5 + 2 = 7 \text{ "}$$
$$5 + 3 = 8 \text{ "}$$
$$\text{and } 5 + 3 + 1 = 9 \text{ "}$$

thus all the ordinal parts of 10 can be weighed. Now suppose a similar example with the tonal system, which will also require five weights, namely, 1, 2, 4, 8, and 10, this is the most natural composition of weights, they are convenient in the operation of weighing and easy for mental calculation.

$$1 = 1 \text{ weight.}$$
$$2 = 2 \text{ weights.}$$
$$2 + 1 = 3 \text{ "}$$
$$4 = 4 \text{ "}$$
$$4 + 1 = 5 \text{ "}$$
$$4 + 2 = 6 \text{ "}$$
$$4 + 2 + 1 = 7 \text{ "}$$
$$8 = 8 \text{ "}$$
$$8 + 1 = 9 \text{ "}$$
$$8 + 2 = 9 \text{ "}$$
$$8 + 2 + 1 = ʓ \text{ "}$$
$$1 + 4 = ƕ \text{ "}$$
$$8 + 4 + 1 = ℰ \text{ "}$$
$$8 + 4 + 2 = ℨ \text{ "}$$
and $8 + 4 + 2 + 1 = ᵳ$ "

thus all the ordinal parts of 10 (16) can be weighed.

It will be observed that the five decimal weights could weigh only the 10th parts of the unit, while the five tonal weights give a nicety of every 16th part; consequently the tonal system has in that case 60 per cent. advantage of the decimal system, and moreover the tonal weights give the natural and desired fractions, quarters, eighths and sixteenths, which is not the case with the decimal weights.

For the natural fractions it will require three more parts to the decimal weights, namely $\frac{1}{2}$, $\frac{1}{4}$ and $\frac{1}{8}$, or expressed by decimals it will be 0·5, 0·25 and 0·125, by which the sixteenth parts can be weighed, but it will be a complicated expression, for instance, 6 parts will be expressed by 0·375 and 7=0·4375, which can never be clearly comprehended, because the mind must be carried away to several thousands for only one figure.

The decimal system can never avoid the expression of the tonal or natural fractions, because they are of daily occurrence in practice, while the tonal system is complete in itself for all uses without exception, and needs no reference to, but will do best without, the decimal system.

If three more parts are added to the five tonal weights, namely, 0·2, 0·4 and 0·8, it can weigh to a nicety of every 128 parts of the unit, the expression will have one decimal (called a tonal) by which the true weight is clearly impressed on the mind.

If you examine all the papers that have been written on the subject in question, including your own, and collect all the advantages and disadvantages of all different systems in your memorandum, then examine well the tonal system, and you will find that all your collected advantages are contained in the tonal system, and all the difficulties and disadvantages are overcome.

Your most humble and
Most obedient servant,
JOHN W. NYSTROM.

On my visit in London I had the pleasure of meeting James Yates, Esq., M. A., F. R. S., Vice President, and Professor Leone Levi, F. S. A. F. S. S., resident Secretary of the International Association.

Mr. James Yates was so kind as to invite me to his house to see the great variety of French metres spoken of in the preceding letters, which was indeed a fine collection. The best form of the metre in the collection, and the one best suited to the artizan I believe is the four folded one. Among the ten folded metres was

85

found what I have before remarked, *none* of the same length, but they differed up to 1½ millimeters. Other forms, parts and divisions of the metre did not however alter my views, but rather strengthened my opinion herein given on the subject.

JOHN W. NYSTROM.

LONDON, *September*, 1860.

———————

—— ——, Esq., *President of the*
—— —— —— *Society, Philadelphia.*

SIR:—I have left in the care of Professor X. a manuscript on a new system of Arithmetic, Weights, Measures and Coins, intended to be submitted to your consideration for publication.

All the engravings and types for the new figures are ready for the press. In it you will find some correspondence with the Decimal Association in London, which is believed worthy of publication for the argument on the French metre.

Yours, most respectfully,
JOHN W. NYSTROM,
1216 *Chestnut St., Phila.*

PHILADELPHIA, *Sep.* 25, 1861.

———————

PHILADELPHIA, *Oct.* 19, 1861.

MR. NYSTROM:

DEAR SIR:—I regret to announce that the report of the Committee on your essay that it recommend that the essay be *not* published, was adopted by the Society at its meeting, last evening; and the MSS. was ordered to be deposited in the archives of the Society, subject to your order.

Professor A. and Professor B. afterwards discussed your tonal system, and Dr. C. the octonal system of Mr. Taylor of this city, whose pamphlet was laid on our table, and seems not to have been noticed by you.* It was suggested that it would be agreeable to publish some abstract account of your system in the running minutes of the proceedings of the meeting.

Very respectfully,
X., *Secretary.*

PHILADELPHIA, *Oct.* 11, 1861.

PROF. X., *Secretary, &c., &c.*

DEAR SIR:—I herewith return the MSS. of Mr. Nystrom, to be examined by the other members of the Committee.

I believe no other report will be necessary than simply to recommend for publication, or the contrary. Although Mr. N.'s papers have failed to convince me of the great gain by substituting the sexta decimal ("tonal") basis of notation, for the decimal, yet it is interesting and instructive, to have such a system fully worked out, and placed before us in all its bearings; that is, when it can be done by a philosophic and competent mind, as is manifested in the case before us. I would, therefore, be in favor of publication, at least as far as page 60, which concludes the main recital. The remainder, which is of equal bulk, is a correspondence between Mr. N. and the officers of the International Decimal Association, at London. To my own apprehension, there is some defect of force and perspicuity

---

* I was not there.—N.

in their criticisms, affording Mr. N. the opportunity of making pretty sharp replies. All this, while it throws light on the subject, and is spicy enough to aid in the digestion, may be considered as somewhat of a repetition.

I have pencilled down a few random comments, and have had them copied on another sheet; and if you please, would like them, with this note, to be handed to Profs. A. and B., along with Mr. N.'s book. I conclude by proposing that the Committee, meet at the hall on the evening of the next meeting of the Society, 18th inst., at a quarter before 8 o'clock, to determine their report.

Very truly, yours,
D.

Hasty comments on Mr. Nystrom's *new basis of Arithmetical Notation.*

(Page 8, *et passim*.) (*First comment*.)

The term "binary division" suggests the necessity for coining a new word. *Binary* refers to a *doubling*, not a halving process. *Demidial* or *dimidiary*, (from dimidium half) would express the very idea; but as yet there is no such word; nor any that expresses the idea. Inasmuch as Mr. N. finds it necessary to make many new *words*, *this one* is respectfully offered.

(Page 21.) (*Second comment*.)

If this new system would afford a relief from *endless fractions*, it would be a triumph over the decimal system; but it does not. While a sixth part is represented in the *decimal* system by ·16666 . . . forever, it stands in the *tonal* system, ·29999 . . . forever. It works well for halves, eighths, sixteenths; but does not work at all for thirds, fifths, sixths, and so on. Yet these divisions are continually occurring in practice.

(Page 40.) (*Third comment.*)

Prices are of every imaginable figure. A car ride is five cents. An exchange ticket seven cents: a pound of sugar, 9, 10 or 11 cents. Mr. N.'s system would be so much bothered by these, that he would probably insist that prices should be such as to make the working easy. He is quite in error about the dollar holding a medium place among the monetary units of the world, and therefore having "a claim to be chosen as a standard." In calling the French franc the smallest unit, he forgets the piastre of Turkey, the rial of Spain, the drachm of Greece, and some others. Nor is the £ sterling the largest unit; there is the milreis of Portugal, and of Brazil. These errors, however, are not material to the merits of the scheme.

(Page 57 to 60.) (*Fourth comment.*)

The account of the Russian *stchoty* or counting machine is interesting. It is essentially the *abacus* of ancient Rome, and of modern China and Japan, the apparatus of a people very low in the scale of mathematical science. Yet Mr. N. would have it brought in our schools and counting houses, to help the new *tonal* system, and "turn the mind from the old basis." It would surely be a retrogade to put away the slate and pencil for this machine.

(Page 69.) (*Fifth comment.*)

If Mr. N. had observed the practice of our market people, he would have found his argument against the decimal system materially weakened. An article costs 38 cents; the buyer hands out a dollar; the seller in making change, is sure to act thus:—first lays down 2 cents, to bring his mind to 40; and then easily makes up the remaining 60 with a half dollar and a dime. So that he first steers for the nearest *ten* to rest upon, and from that completes the operation. He never thinks of mentally subtracting 38 from 100, unless he be an old accountant, or schoolmaster.

It may be observed that Mr. Alfred B. Taylor of this city, constructed an ingenious system on the *octonal* basis. Mr. Pitman, the celebrated phonographer, urged a *duodecimal* reform; and Dr. Patterson used to mourn that our arithmetic was not based upon 12 instead of 10.

Philadelphia, 1216 *Chestnut street, Oct.* 23, 1861.

PROFESSOR X., *Secretary &c., &c.*

DEAR SIR:—Your favor of the 19th inst. is at hand, I am sorry to hear the —— Society did not deem my manuscript worthy of publication. It is true I have not noticed Mr. Taylor's *octonal* system, as my *tonal* system was written in Russia long before Mr. Taylor's *octonal* system was published in America, and even if I had seen it, it would not have altered one sentiment in my manuscript. The —— Society "suggested that it would be agreeable to publish some abstracts," which I suppose from Mr. D.'s letter to you dated Oct. 11th, would be to omit my correspondence with the International Decimal Association in London. When my Tonal system is published, I shall omit nothing of the manuscript, even my correspondence with and remarks made by the —— Society will be published, as I desire to have the subject thoroughly ventilated.

It may be found that there " are some defects of force and perspicuity in" the comments made by Mr. D. which affords me a second " opportunity of making " a pretty sharp reply."

The first comment is " The term binary division sug- " gests the necessity of coining a new word." I have only to refer to page 61, where the International Decimal Association in London, uses the same expression. " Binary" can be applied to halving as well as doubling, the difference is only to go up or down the steps. The word " Binary Divisions" is freely used in Mr. Taylor's report on the octonal system. A binary compound, say chloride of sodium, Na, Cl, contains half of each substance, which is an example of a binay halving process.

Second comment. " If this new system would afford " a relief from *endless fractions*, it would be a triumph " over the decimal system, but it does not." It would indeed be a triumph! but Mr. D. will never be satisfied on that point, for let us even propose one system of arithmetic for each fraction, or attempt to invent a system of arithmetic that would have no prime numbers, he will still be disappointed. Can Mr. D. describe a circle through these four points .:·? and it will be a triumph in geometry. " While a sixth part is repre- " sented in the decimal system by 0·16666, forever, it " stands in the tonal system 0·29999, forever. It works " well for halves, eighths, sixteenths; but does not work " at all for thirds, fifths, sixths, and so on, yet these " divisions are continually occurring in practice." Fifths are generally employed for the necessity of accommodating the decimal system which imposes so much inconvenience upon us. Sixths and 12ths are often used in practice as an improvement on, or to avoid 5ths and 10ths; with the exception for the circle, 6ths and 12ths are of little importance compared with the binary fractions. In the following table are set down the fractions in question, with one, two, and three, decimals with their errors, in the Tonal, Decimal, and Octonal systems.

Systems.	One Decimal.		Two Decimals.		Three Decimals.	
	Fraction.	Error.	Fraction.	Error.	Fraction.	Error.
Tonal,	$\frac{1}{6} = 0\cdot2$	0·0416	$\frac{1}{6} = 0\cdot09$	0·0026	$\frac{1}{6} = 0\cdot299$	0·00016
Decimal,	$\frac{1}{6} = 0\cdot1$	0·0666	$\frac{1}{6} = 0\cdot16$	0·0066	$\frac{1}{6} = 0\cdot166$	0·00066
Octonal,	$\frac{1}{6} = 0\cdot1$	0·0416	$\frac{1}{6} = 0\cdot12$	0·0104	$\frac{1}{6} = 0\cdot125$	0·00065
Tonal,	$\frac{1}{3} = 0\cdot5$	0·0208	$\frac{1}{3} = 0\cdot55$	0·0013	$\frac{1}{3} = 0\cdot555$	0·00008
Decimal,	$\frac{1}{3} = 0\cdot3$	0·0333	$\frac{1}{3} = 0\cdot33$	0·0033	$\frac{1}{3} = 0\cdot333$	0·00033
Octonal,	$\frac{1}{3} = 0\cdot2$	0·0830	$\frac{1}{3} = 0\cdot25$	0·0280	$\frac{1}{3} = 0\cdot252$	0·00130

It will be seen in this table that the fraction $\frac{1}{6}$ expressed by one decimal has by the tonal system 60 per cent advantage in the correctness over the decimal system. With two decimals 250, and with three decimals 410 per cent. advantage in the correctness. Still Mr. D. says these fractions "does not work at all." For thirds the favor is still greater for the tonal system.

Third Comment. "Prices are of every imaginable "figure. A car ride is five cents, an extra ticket 7 "cents." In my manuscript I speak about omnibus prices, as they were when I left America for Europe in the spring of 1856, which was written in Russia in the year 1859, when I knew nothing about the street railroad arrangement. Six cents or rather $6\frac{1}{4}$ is a very general price for articles, as being $\frac{1}{16}$th part of a dollar. After my return from Europe, I made the following observations on car ride prices:

Frankford and Southwark Passenger R. R. Co.

FARES.

Southwark to Front and York streets,	5 Cts.
" Frankford,	10 "
Germantown road to Frankford,	7 "
Southwark to Episcopal Hospital,	7 "
Berks street to Harrowgate,	5 "
Frankford to Hart lane,	5 "
For Children under 12 years,	3 "

By examining this price table we find that all the prices are not only odd but of prime numbers 3, 5 and 7. The base price for a car ride is 5 cents, and for long distances double price, 10 cents. For intermediate distances such as from Germantown to Frankford, and from Southwark to the Episcopal Hospital, is charged 7 cents, showing an attempt to charge a price half way between 5 and 10, which should be $7\frac{1}{2}$ cents, but our coins as well as our decimal base does not permit such division, for which we must be contented with the prime number 7. It is a general custom over the world, to charge half price for children, which in this case should be $2\frac{1}{2}$ cents, but as we have no such coins, it is made up to 3 cents.

Let us now see what the *tonal* prices would be. By the tonal system it is very likely that the base price for a car ride would be 1 shilling, ($6\frac{1}{4}$ cents,) but suppose even this to be too high, and the exact value of 5 cents is required, which would be 𝒱 tonal cents, the price table would be as follows:

	Decimal.	Tonal Prices.	
Southwark to York Street,	5 cts.	𝒱 cts.	1 s.
" Frankford,	10	1·8 s.	2
Germantown rd. to Frankf'd,	7	1·2	1·8
Berk street to Harrowgate,	5	𝒱 cts.	1
Children half price,	3	6	8 cts.

The tonal price in both cases are all of even and easy countable numbers, and divides the prices as desired in practice.

The Sanford's Opera bill says:

  Admittance, - - - 25 cents.
  Half price for children, - 13 "

Pennsylvania Railroad trains leave Philadelphia at:

	Decimal.	Tonal.
Mail train,	8 A. M.	5·8 T.
Fast line,	11·30 A. M.	7·2
Through express,	10·30 P. M.	6
Harrisburg train,	2.30 P. M.	9·9

In this time table there is no confusion of A. M. and P. M. in the tonal column, but the correct time is expressed by the fewest possible numbers, clear to the mind at the first glance. T means the hour mark.

The original meaning of A. M. and P. M. is not generally known further than that it means forenoon and afternoon, but even that is sometimes confused.

Two coal-miners, Jack and Harry, arrived at a railway station in England, and examined the time table, when the following conversation took place:

Jack. I say Harry, what does P. M. mean?

Harry. Penny a mile, to be sure.

Jack. What does A. M. mean, then?

Harry. Oh, that must be a A penny a mile.

Jack. Then we will go by the A. M. line.

Jack and Harry were not acquainted with *ante meridiem* and *post meridiem*.

In the crowded railway guides it is often difficult to find out whether a noted time means in the forenoon or afternon, and it is often necessary in tables to leave a separate column for A. M. and P. M.

Rev. E. Burnham preaches in the Concert Hall, Philadelphia, every Sunday at:

$10\frac{1}{2}$ A. M., 3 P. M., and $7\frac{1}{4}$ evening.
Tonal time, 7   9   8 T.

Our present system employs *twelve* characters and one word, where the tonal system uses only *three* characters and the hour mark. The Rev. gentlemen seems disposed to divide his time as in the *tonal system*.

Third Commemt. "Mr. Nystrom's system would be so "much bothered by those that he would probably insist "that prices should be such as to make the working easy."

Every shop keeper attempts to arrange his prices into easy countable figures, but the inconvenience of the decimal system is so great that it is difficult or rather impossible to satisfactorily attain that object, as is readily seen in store windows by prices marked on articles $37\frac{1}{2}$ cents, $81\frac{1}{4}$ cents, &c., &c., an attempt to approach the tonal system. If Mr. D. will take the trouble to examine the shop practice, he may discover that prices are already arranged to suit the tonal system, in spite of the *bothered* decimal coins.

Fourth Comment. "He is quite in error about the "dollar holding a medium place among the monetary "units of the world." How does Mr. D. know that the dollar is not holding a medium place, among monetary units? Did he try it? And if so, why not favor us with his result? Is it more or less? The following table contains the present monetary unit in most parts of the world.

The table can however be varied by judgment of different units existing in different countries, that any one who feels disposed to make comments on it

95

has an extensive field to operate upon. When I first worked out this medium monetary unit the result came much nearer the dollar than in this table.

Monetary units in most parts of the world.

	$. cts.		$. cts.
South American & Mexican *dollar*,	1·05	Spanish, *Piaster*,	1·06
		Sweden, *Riksdaler*,	0·27
Chili and New Granada *dollar*,	0·96	Turkey, *Piastre*,	0·04
		Wirtemberg, *Thaler*,	1·00
China *dollar*,	1·43	Total,	18·86
Austria, Bohemia & Bavaria, *Florin*,	0·50	Divided by 20 will be 94 cents as a medium unit.	
Denmark, *Specidaler*,	0·96	The main money of Europe is Pound Sterling,	4·86
France, Belgium, Switzerland and Italy, *Franc*,	0·18		
		Florin of Germany,	0·45
Great Britain, £ *Sterling*,	4·86	Thaler of Germany,	0·70
		Ruble of Russia,	0·77
India, *Rupies*,	0·46	Franc of France,	0·18
Hamburg, *Mark*,	0·30	Riksdaler of Sweden,	0·27
Hanover, *Thaler*,	1·10		7·23
Holland, *Florin*,	0·41	7·23 : 6 = 1·20	
Prussian, *Thaler*,	0·72	0·94	
Portugal, *Millrea*,	1·12	2 ⌐ 2·14 ⌐ 1·07 dols.	
Rome, *Scudo*,	1·04	Which can be considered a medium monetary unit of the world.	
Russian, *Ruble*,	0·77		
Saxony, *Thaler*,	0·63		

Fifth Comment. "In calling the franc the smallest "unit he forgets the piaster of Turkey, the real of Spain, "the drachm of Greece, and some others. Nor is the "£ sterling the largest unit; there is medries of

"Portugal, ($1 12) and of Brazil." I beg to assure Mr D. that the units referred to are not forgotten.

If we go to those extremes it is difficult to know where to begin or end. We may call the American eagle 10 dollars a unit, the Napoleon 20 franc, Imperial 100 fr, doubloon of Spain, Central and South America, about 15 dollars. Russian imperial 5·15 Rubles. Dobras of Brazil 34 dollars, and other monetay units ranging from a fraction of a cent towards 50 dollars. If we go further to the corners of the land among the Esquimaux's or the Calmucks we may find a *hide* of any animals or a *heap of hay* to be a unit for trading. The £ sterling and franc are the extreme units known and handled over the whole world, in preference of which the outside units of Greece and of the Esquimaux could not be admitted in a general statement. The French Franc is used in Belgium, Switzerland, Italy and Algiers, by a population of about 70 million; the £ sterling is used by a population of I suppose 50 millions.

The franc and £ sterling put together, I believe would exceed all the rest of the money in the world. Mr. D. says "These errors however are not material to the merits of the scheme." I have not been able to find a single expression in Mr. D.'s comments that have any bearing whatever on merit or folly of the tonal system.

Sixth Comment. "Yet Mr. Nystrom would have it
" (the counting machine) introduced into schools and
" counting houses to help the new tonal system and
" turn the mind from the old basis. It would surely
" be a retrograde to put away the slate and pencil for
" this machine." I would like very much to see Mr.

D. with a slate and pencil alongside a Russian with a tshoty, I would give the example for calculation, and Mr. D. would soon find the utility of the Russian tshoty, and the retrograde of his slate and pencil. When the tonal system is well-acquired, the counting machine would be found superfluous, but for the first acquirement the slate and pencil would not answer the same purpose as the counting machine.

Seventh Comment. " If Mr. Nystrom had observed " the practice of our market people, he would have " found his argument against the decimal system " materially weakened." After having read Mr. D.'s hasty comments in the Archives of the —— Society, I proceeded up Chestnut street, when opposite the Masonic Hall I observed at the northwest corner of Eighth and Chestnut, at the doorway of Sharpless' store, a pile of dry-goods upon which was a paper sign marked with figures as big as my hat, " Extra quality $87\frac{1}{2}$ cts." (per yd.) Arriving at the southeast corner of Eighth and Chestnut streets, I saw at the Eighth street doorway of the same store, two piles of dry-goods marked one $37\frac{1}{2}$ cts. and the other $62\frac{1}{2}$ cts. (per yard). Looking up Eighth street I saw at the northeast corner of Eighth and Zane streets, in the doorway of a store, a crinoline in the inside of which was a paper sign marked $37\frac{1}{2}$ cts. I walked up to this store where I found in the window fifteen articles marked with the following prices, $6\frac{1}{4}$ cts. $12\frac{1}{2}$ cts. $18\frac{3}{4}$ cts. 25 cts. $31\frac{1}{4}$ cts. $37\frac{1}{2}$ cts. &c., &c., all arranged to accommodate the easy counting as in the tonal system.

Returned and went into Mitchell's restaurant, 808 Chestnut street, where I took a cup of coffee and cakes; was handed an ivory ticket upon which was engraved the number 19, indicating the price of my refreshment. At the counter I inquired why they charged the odd number 19 cts. and was answered "It ought to be 18¾ cents, but as we have no such "coins we make it even to 19." I then asked, would not 20 be a more even number? And was answered, "20 is not even in a dollar." Here you will find that the prime number 19 is called even, and the even number 20 is considered to be odd.

It is literally true, that 20 is odd in 100, and that odd numbers as 5, 25 and 75, are even in 100.

Upon further inquiry of their prices, I was shown to the other side in the store, to a box of about 13 inches square, divided into 36 compartments, each containing ivory tickets marked with the following prices

Mitchell's Price Ticket Box.

6¼ct.	13	19	25	31	38
44	50	56	63	69	75
81	88	94	100	106	112
118	125	131	137	144	150
162	168	175	181	200	275
281	287	300	306	312	318

Here you will find that it is attempted to arrange the prices to suit the dollar divided into 16 parts, as in the tonal system. You will observe that most of the prices are of odd and prime numbers, and to make them perfectly correct, most of them ought to be accompanied with fractions. It is evident that those prices are considered easy to the mind in the market, and how much better would it not be, if our arithmetic was based on the same principle; every coin proposed in my tonal system agrees correctly with those prices.

Taylor's saloon in New York, and a great many other establishments, have similar arrangements of prices.

I left the restaurant, walked up Chestnut street, stopped at the store of Le Boutillier Brothers, No. 912, where I found prices marked in the window $62\frac{1}{2}$ cents, $87\frac{1}{2}$ cents, 75 cents, &c., &c.

At Besson & Son's mourning store, 918 Chestnut street, I found in the window marked the following prices.

De Laines,	$12\frac{1}{2}$ cents.	Reps Anglais,	$37\frac{1}{2}$ cents.
Cravellas,	25 cents.	Mousselin,	$6\frac{1}{4}$ cents.
De Laines,	$18\frac{3}{4}$ cents.	Other articles,	$62\frac{1}{2}$ cents.
Grandrill,	$31\frac{1}{4}$ cents.	One article,	44 cents.

No price had any indication of decimal division. Went home to my house 1216 Chestnut street, where I pay for my washing $62\frac{1}{2}$ cents per dozen. About two weeks ago, I was charged by a shoemaker $37\frac{1}{2}$ cents for mending a pair of boots.

Will Mr. D. yet think that my observation of market practice would materially weaken my argument against the decimal system? and in case he suppose that this is my first attention to market practice, I beg to remind him that in my manuscript it is plainly stated that it is my observation of the inconvenience of the decimal system and arithmetic in the shop and market that has led me to propose the tonal system. In Mr. D.'s argument on the 38 cents, he still brings the mind to the high numbers of 40, 60 and 100, where the tonal system would bring it only to the base 10. The price 38 cents would be 6 shillings *tonal*.

The following table contains the most common market price.

Market Prices in Cents.	Tonal Shillings or 16ths of a Dollar.	Nearest Cents as Mitchell's Tickets.
$6\frac{1}{4}$	1	6
$12\frac{1}{2}$	2	12 or 13
$18\frac{3}{4}$	3	19
25	4	25
$31\frac{1}{4}$	5	31
$37\frac{1}{2}$	6	37 or 38
$43\frac{3}{4}$	7	44
50	8	50
$56\frac{1}{4}$	9	56
$62\frac{1}{2}$	10	62 or 63
$68\frac{3}{4}$	11	69
75	12	75
$81\frac{1}{4}$	13	81
$87\frac{1}{2}$	14	87 or 88
$93\frac{3}{4}$	15	94
100	16	100

Can Mr. D. discover any utility in the centre column of this table, compared with the two outside ones?

Lastly, Mr. D. says: "It may be observed that "Mr. Alfred Taylor of this city, lately constructed an "ingenious system on the octonal basis. Mr. Pitman, "the celebrated phonographer used a *duodecimal* reform; "Dr. Patterson used to mourn that our arithmetic was "not based upon 12, instead of 10." It seems from these statements, that Mr. D. has no preference to any one of the three basis, 8, 12 and 16, that if I had proposed 14 as a base, he may have given it the same consideration; and if such is the case with the ——— Society, I do not wonder at all, that the tonal system was rejected for publication.

There is nothing new in merely proposing a better base for our arithmetic; that I believe has been done since the time of Charles XII., of Sweden, by hundreds, and been thought of by thousands; for any self thinker with good reason of mind, sees plainly the *folly of our decimal arithmetic*. I am surprised to find so many of the first leaders of the scientific world to be so short sighted, as not to see the inconveniences, but propagates a system so unnatural in all its bearings.

About a year ago there was an Italian in London, who proposed the *duodecimal* system, but in no case have I found any of such systems worked out with examples into a practical shape. Most of the propositions have been made by mere scientific men, who have given excellent accounts of the history of arithmetic, and finished by merely proposing a better base. I believe myself to have commenced and continued from where they ended, and I suppose you to know what they have said.

Purely scientific men are not the proper persons to handle this practical subject, for the decimal arithmetic is so clear to them, that they manage the figures and come to their results as easy as a musician who plays the crank organ. Their lack of direct application of their science to practice, screens away the real inconvenience of our decimal arithmetic, which is readily proved by feeble remarks frequently made by such men. Many of them confine themselves more to style of language than to the *substance* of the subject. It is not sufficient merely to propose or say that 8, 12 or 16, would be better as a base, but in order to make a clear and correct impression of its utility, it is necessary to enter into details with examples, that any one may be able to estimate its advantages without taxing his own mind. Still the nature of the subject is such, as to be apt to be called *curious* at the first glance.

The *octonal* system has two serious objections:

First. That the base 8 is too small. Our experience with the decimal arithmetic is, that 10 is too small as a base.

Secondly. As we progress in this world, generation after generation, we require larger and larger numbers in our transactions. That which Moses counted by thousands, are by us counted by millions, and I venture to say, that with the present decimal arithmetic, there are very few who have a clear conception of the immense number of one million; and the more complication we have to lead us to such a number, the more cloudy it will be to the mind. The immense numbers necessary in astronomy, expressed by decimal arithmetic are inconceivable, while the *tonal* system gradually leads the mind towards infinitum.

In the *octonal* system we have two figures already at 8, three at 64, and four at 512; while in the *tonal* system we have two figures first at 16, three at 256, and four at 4096. One *decimal* million expressed by *octonal* arithmetic will be 3,641,100, and by *tonal* arithmetic 94,240, which is a difference of two figures. Also for *decimal* fractions the *octonal* requires more figures than the *tonal* system for the same nicety. The *duodecimal* has many advantages over the *decimal* system, particularly in thirds and sixths, but this is overbalanced by the serious objection of it not admitting binary division to infinitum. Sixths and thirds work much better in the *tonal* than in the *decimal* system, as seen in the table, page 91.

Many self-thinkers express their regrets that the arithmetical system was not from the beginning founded on a better base. Many of them, I believe, prefer the duodecimal system, and some express their wish that man would have had six fingers on each hand, which might have led to a duodecimal system. The Sixdiopt Family, in Central America, have six fingers on each hand, and six toes on each foot; they might have had accomplished that object. By this theory, I would, of course, prefer eight fingers on each hand. I know no tribe of people that can accommodate me, but am satisfied that five fingers will answer for the tonal system. Should we now succeed to introduce a duodecimal system, *our descendants would surely wish for the tonal system of us, as we wish for a duodecimal system from our ancestors.*

If the Arabic notation employed in our present *decimal* arithmetic had been suggested to Moses when he wrote the ten commandments on Mount Sinai, he would surely have made similar remarks as that made on the *tonal* system, that such *curious looking characters* could not be understood by his people.

My manuscript on the *tonal* system has been sent to a great many places for publication. The Franklin Institute thought it would have a very serious effect on the number of subscribers of their journal! The Smithsonian Institute would not publish it, because they had so much of the same kind before!!! The U. S. Coast Survey stated they would publish it if recommended by a member of Congress.* And lately the —— Society of Philadelphia has rejected it, perhaps on the remarks herein replied to, but Mr. D. recommended its publication.

I return herewith the pamphlet on Mr. Taylor's *octonal* system, and thank you very much for calling my attention to it. It is indeed an interesting and ably written work. I suppose I must follow the track of Mr. Taylor, and go to Boston to get my *tonal* system understood and appreciated.

The heading of this letter is dated October 23d, when I intended to write but a few lines, but when I got into it, I could not well cut it off until it reached nearly twenty-four pages. It is now the 28th of October.

Your humble and obedient servant,

JOHN W. NYSTROM.

---

* The idea of showing to a member of Congress this manuscript and calculations, with t's, i's, &c., among the figures! he would surely pronounce me a funny fellow. When scientific men, as at the Franklin and Smithsonian Institutes, ———— Society, and others, cannot appreciate the subject, what can we then expect of a member of Congress?

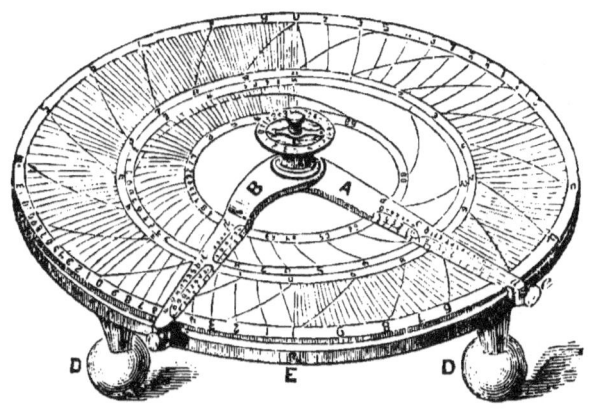

NYSTROM'S CALCULATOR.

This calculating machine consists of a silvered brass plate of about nine inches in diameter, on which are fixed two movable arms, extending from the centre to the periphery. On the plate are engraved a number of curved lines in such form and divisions that with their intersection with the arms, the most complicated calculations can be performed almost instantly.

The arrangement for trigonometrical calculations is such that it is not necessary to notice the functions *sine, cosine, tangent,* &c., operating only by the angle expressed in degress and minutes, and without any tables, which makes it so easy that any one who can read figures, will be able to solve trigonometrical questions. Any kind of calculation can be performed on this instrument, no matter how complicated it may be, whilst there is nothing intricate in its use. The author, who is the inventor of the calculator, has thoroughly tested its practical utility. All the calculations in Nystrom's Pocket Book of Mechanics and Engineering have been computed by this instrument.

Teachers are generally dependent upon text book for examples, when it is easy for the pupil, knowing where it comes from, to be furnished with an answer; but with the calculator, the teacher can vary the examples *ad libitum*, and the answer is almost instantly at hand, while the pupil is thrown on his own resources for the proper solution, and his real acquirement is tested.

The price of the Calculator, with complete description and examples how to use it, $20.

Manufactured by Wm. J. Young, 43 North Seventh Street. Sold by James W. Queen & Co., 924 Chestnut Street, Philadelphia.

NYSTROM'S
Pocket Book of Mechanics and Engineering

PUBLISHED BY

J. B. LIPPINCOTT & CO.
PHILADELPHIA.

TRUBNER & CO., LONDON.

This Pocket Book is now in its fifth edition, revised and enlarged with fifty new pages, and contains new and original matter not to be found in other Engineering books. Price, $1.50.

www.ingramcontent.com/pod-product-compliance
Lightning Source LLC
Chambersburg PA
CBHW020154170426
43199CB00010B/1038